Recent Titles in This Series

(Continued in the back of this publication)

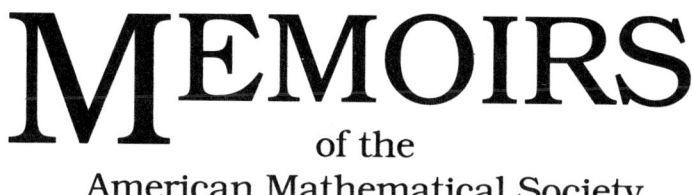

MEMOIRS
of the
American Mathematical Society

Number 547

On the Classification of C*-algebras of Real Rank Zero: Inductive Limits of Matrix Algebras over Non-Hausdorff Graphs

Hongbing Su

March 1995 • Volume 114 • Number 547 (third of 4 numbers) • ISSN 0065-9266

American Mathematical Society
Providence, Rhode Island

1991 *Mathematics Subject Classification.*
Primary 46L05.

Library of Congress Cataloging-in-Publication Data

Su, Hongbing, 1957–
   On the classification of C^*-algebras of real rank zero : inductive limits of matrix algebras over non-Hausdorff graphs / Hongbing Su.
       p.   cm. – (Memoirs of the American Mathematical Society, ISSN 0065-9266; no. 547)
   "March 1995, volume 114."
   Includes bibliographical references and index.
   ISBN 0-8218-2607-7
   1. C^*-algebras.   2. K-theory.   I. Title.   II. Series.
QA3.A57   no. 547
[QA326]
510 s–dc20
[512′.55]   94-43209
                                                                                         CIP

Memoirs of the American Mathematical Society

This journal is devoted entirely to research in pure and applied mathematics.

Subscription information. The 1995 subscription begins with Number 541 and consists of six mailings, each containing one or more numbers. Subscription prices for 1995 are $369 list, $295 institutional member. A late charge of 10% of the subscription price will be imposed on orders received from nonmembers after January 1 of the subscription year. Subscribers outside the United States and India must pay a postage surcharge of $25; subscribers in India must pay a postage surcharge of $43. Expedited delivery to destinations in North America $30; elsewhere $92. Each number may be ordered separately; *please specify number* when ordering an individual number. For prices and titles of recently released numbers, see the New Publications sections of the *Notices of the American Mathematical Society.*

   Back number information. For back issues see the *AMS Catalog of Publications.*

   Subscriptions and orders should be addressed to the American Mathematical Society, P. O. Box 5904, Boston, MA 02206-5904. *All orders must be accompanied by payment.* Other correspondence should be addressed to Box 6248, Providence, RI 02940-6248.

Memoirs of the American Mathematical Society is published bimonthly (each volume consisting usually of more than one number) by the American Mathematical Society at 201 Charles Street, Providence, RI 02904-2213. Second-class postage paid at Providence, Rhode Island. Postmaster: Send address changes to Memoirs, American Mathematical Society, P. O. Box 6248, Providence, RI 02940-6248.

10 9 8 7 6 5 4 3 2 1   00 99 98 97 96 95

Contents

Abstract

In this paper a K-theoretic classification is given of the real rank zero C*-algebras that can be expressed as inductive limits of sequences of finite direct sums of matrix algebras over finite connected graphs (possibly with multiple vertices). The special case that the graphs are circles is due to Elliott.

Key words and phrases: K-theory, classification, C*-algebras, inductive limits, real rank zero. AMS subject classification number (1985 version): Primary: 46L05.

To My Family

Chapter 1. Introduction

1.1. The problem we are interested in is to classify certain C*-algebras by their K-theory data. Given a C*-algebra A there are two abelian groups associated with it. Namely, $K_0(A)$, which is often an ordered group, and $K_1(A)$ [2]. The complete invariant in the present classification will be the graded group $K_* = K_0 \oplus K_1$, together with the distinguished subset consisting of the pairs $([e], [u])$, where e is a projection in the algebra and u is a partial unitary with support (and range) e [11]. Following [11], we shall call this set the graded dimension range.

The first result of this kind was the classification of inductive limits of sequences of finite direct sums of matrix algebras (called AF algebras), in which case K_1 is trivial [10]. The group K_0, together with the dimension range, is a complete invariant. In [11], this classification was extended. It was shown that if one replaces the matrix algebras by matrix algebras over the unit circle and restricts the limit C*-algebras to have real rank zero (see Definition 1.3), then K_*, as a graded group, together with the graded dimension range, is a complete invariant. Recently, it was shown in [15] that this class of C*-algebras contains all the irrational rotation C*-algebras (also see [8]).

In this paper the K-theoretic classification theorems are extended to the real rank zero C*-algebras that can be expressed as inductive limits of finite direct sums of matrix algebras over finite non-Hausdorff graphs (Theorem 8.3).

In [11] G. A. Elliott asked the following question: is K-theory data a complete invariant for the class of separable nuclear C*-algebras of real rank zero and stable rank one (i.e., such that the invertible elements are dense)? A positive answer would be a very important result. One way to attack the problem is to express the C*-algebras in question as inductive limits. An important step in this direction is the result that irrational rotation algebras are inductive limits [15]. Another

This research was supported by a Ontario Graduate Scholarship, a University of Toronto Special Fellowship, and the financial assistance of the Department of Mathematics, University of Toronto. The author wishes to thank G. A. Elliott for suggesting the topic of this paper and for many helpful discussions. He also wishes to thank M.-D. Choi, T. Giordano and L. T. Gardner for their help.

Received by the editors March 20, 1993

way to study the problem is to generalize the inductive limit classification theorem. This is one of the motivations for us to consider C*-algebras of this kind.

Another reason to consider C*-algebras of this form is to study group actions. Recently, a number of people have constructed interesting group actions on C*-algebras of this type. And this has answered some difficult questions (see 1.3). In [2], it was shown that the crossed product of $\mathbb{Z}_2$ with $M_n(C(T))$ via the flip on the unit circle T is M_{2n} over the unit interval with multiple end points. These results suggested the problem of classifying the real rank zero C*-algebras that can be expressed as inductive limits of finite direct sums of matrix algebras over arbitrary finite graphs with multiple vertices.

Inductive limits of this type (with or without the real rank zero property) have been studied for different purposes ([9] [15] [16] [17] [23] [25]). A summary can be found in [14].

The K-groups for the C*-algebras considered above are torsion free. In [11], it was shown that K_*, together with the dimension range, continues to classify the real rank zero C*-algebras arising as inductive limits of finite direct sums of matrix algebras over the unit interval with dimension drops. This class of C*-algebras provides torsion groups for K_1. In [24], the classification theorem is extended to the real rank zero C*-algebras that can be expressed as inductive limits of finite direct sums of matrix algebras over finite graphs with dimension drops at their vertices.

1.2.   We shall consider real rank zero C*-algebras that can be expressed as inductive limits of sequences of finite direct sums of basic building blocks defined as follows. For simplicity, we will assume that both the C^*-algebras and the connecting maps are unital.

Definition.   Let X be a finite connected graph and denote by $C(X)$ the C*-algebra of complex-valued continuous functions on X. A C*-algebra A will be said to be a basic building block if A is isomorphic to a sub-C*-algebra of $C(X) \otimes M_n$ of the following form:

$$\{f \in C(X) \otimes M_n \mid f \text{ has block diagonal form at each vertex of } X\}$$

For some positive integer n. The number of blocks at each vertex is called the multiplicity of that vertex.

In this paper, a basic building block as in the definition will be said to have multiple vertices spectrum X, or simply spectrum X. If x is a vertex of X and the fibre of A at x has k blocks, then we may write $x = \{x_i\}_{i=1}^k$ where each x_i corresponds to a block. We may call those x_i vertex points.

As mentioned in 1.1, in order to get torsion groups for K_1, one may consider dimension drop basic building blocks ([11], [24]). To define a basic building block with dimension drop, one still considers a sub-C*-algebra of $M_n(C(X))$ (see definition above). Instead of requiring the fibre at each vertex to have arbitrary diagonal block form, one requires the fibres at some vertex of X to be $M_{n/m}$ for some m dividing n, i.e., the fibre at such a vertex is m copies of $M_{n/m}$ along the diagonal. The K_1 group of this C*-algebra may have torsion component $\mathbb{Z}_m$.

1.3 Definition. *A unital C*-algebra is said to be of real rank zero if the set of invertible self-adjoint elements is dense in the set of all self-adjoint elements [7].*

As we mentioned in 1.1 real rank zero C*-algebras that can be expressed as inductive limits of finite direct sums of basic building blocks contain AF algebras and the irrational rotation algebras, a non-trivial result obtained in [15]. These C*-algebras are also important in the study of group actions.

In [5], it was shown that for any finite group G, there exists an AF algebra A and an action α on A such that the fixed point algebra A^{α} is not AF. In the proof, A was constructed as inductive limit of basic building blocks with multiple end point intervals as their spectra. In this way, actions other than inner were constructed on basic building blocks. The result for the group of order two was obtained in [3] and [21].

In [2], it was shown that the crossed product of continuous functions on the unit circle $\mathbb{T}$ with $\mathbb{Z}_2$ via the flip $z \to \bar{z}$ on $\mathbb{T}$ is M_2 over the unit interval with multiple ends.

These results suggested the problem of classifying the C*-algebras of this kind as well as the inductive limits of finite group actions.

1.4. Outline

As we mentioned in 1.1, the objective of this work was to show that the K-theory data of the C*-algebras considered in 1.2 is a complete invariant (Theorem 8.3). This paper can be divided into three parts. The first part contains chapter 2, chapter 3, chapter 4 and chapter 5. The main result of this part is Theorem 5.5. It says that for a C*- algebra of the kind considered in 1.2, one can replace arbitrary finite graphs by certain special graphs. The second part contains chapter 6, chapter 7 and chapter 8. The main result of this part is the K-theoretic classification (Theorem 8.3). Although the main supporting results are the existence and uniqueness theorems from Chapter 6 and Chapter 7, the theorems from the previous chapters are needed. The final chapter contains some applications.

Chapter 2.   Small Spectrum Variation

In this section we develop a technical lemma which will be used later. Then we will prove a small spectrum variation theorem for real rank zero inductive limit C^*-algebras. The following lemma 2.1 is a combinatorial result (see [19]).

2.1 Lemma.   *Let $\{x_i\}_{i=1}^n$ and $\{y_i\}_{i=1}^n$ be two groups of points in a metric space with distance function $d(\cdot,\cdot)$. For a fixed integer ℓ and for each i, let $X_i = \{x_i, \ldots, x_i\}$ and $Y_i = \{y_i, \ldots, y_i\}$, where the cardinality $|X_i| = |Y_i| = \ell$. Given $\varepsilon > 0$ and let $Y^{(i)} = \{y_j \in Y_j \mid d(x_i, y_j) < \varepsilon\}$ for $i = 1, 2, \ldots, n$. Then the following two conditions are equivalent:*

(a)   *For any distinct k indexes $\{i_t\}_{t=1}^k$ with $1 \le k \le n$,*

$$\left| \bigcup_{t=1}^k Y^{(i_t)} \right| \ge \ell k$$

where $\left| \bigcup_{t=1}^k Y^{(i_t)} \right|$ is the cardinality of $| \bigcup_{t=1}^k Y^{(i_t)} |$, counting multiplicity

(b)   *There exists a pairing between $\{x_i\}_{i=1}^n$ and $\{y_i\}_{i=1}^n$ such that the distance between the two points in each pair is to within ε.*

2.2.   Let X be a compact metric space with distance function $d(\cdot,\cdot)$ and let $M \ge 1$ be a fixed number. For any closed subset $S \subset X$, we are going to define a function $K_{S,M}$ in $C(X)$ such that $K_{S,M} = 1$ on S and $K_{S,M} = 0$ outside $\frac{1}{M}$ neighbourhood of S. In between, $K_{S,M}$ is piecewise linear with slope M.

First, for any $x \in X$, we define a "cone" function $K_{x,M}$ in $C(X)$ as follows:

$$K_{x,M}(t) = \begin{cases} 1 & t = x \\ 0 & t \in \{t \mid d(t,x) > \frac{1}{M}\} \\ 1 - Md(x,t) & t \in \{t \mid d(t,x) \le \frac{1}{M}\} \end{cases}$$

For a closed subset $S \subset X$, define

$$K_{S,M}(t) = \max\{K_{x,M}(t) \mid x \in S\}.$$

4

It is easy to see that $K_{S,M}$ belongs to $C(X)$. Furthermore, $K_{S,M} = 1$ on S and $K_{S,M} = 0$ outside a $\frac{1}{M}$ neighbourhood of S. We will sometimes simply write K_x and K_S.

For a positive integer ℓ, $K_{S,M} \otimes I_\ell$ is a positive element in $M_\ell(C(X))$. We will denote it by $h_{S,M,\ell}$, or simply h_S, and call it a test function. Use these notations we have the following

Lemma. Let X be a compact space with distance function $d(\cdot, \cdot)$ and let $\{x_i\}_{i=1}^n$ and $\{y_i\}_{i=1}^n$ be two groups of points in X. Fix $1 > \varepsilon > 0$ and assume that the eigenvalues of $\{h_S(x_i)\}_{i=1}^n$ and $\{h_S(y_i)\}_{i=1}^n$ are within ε one by one, in increasing order, counting multiplicity, for each test function h_S associated with a fixed positive number $M \geq 1$; then $\{x_i\}_{i=1}^n$ and $\{y_i\}_{i=1}^n$ can be paired to within 3ε one by one.

Proof: For any $x_{n_1}, \ldots, x_{n_k}$, let $S = \{x_{n_1}, \ldots, x_{n_k}\}$. $S \subset X$ is closed. Now the assumption in the lemma gives condition (a) of Lemma 2.1. This completes the proof of the lemma.

2.3. Since there are infinitely many closed subsets in X, there are infinitely many test functions in $C(X) \otimes M_\ell$ associated with a fixed number M. The next lemma deals with this problem.

Lemma. Let X be a compact metric space with distance function $d(\cdot, \cdot)$ and let $H \subset M_\ell(C(X))$ be all the test functions associated with a positive $M \geq 1$. For given $0 < \delta < \frac{1}{2M}$, there exist finitely many closed δ-balls, say $S_1, \ldots, S_n$, such that $\{h_S \mid S = \bigcup_{t=1}^k S_{i_t}, \{i_t\}_{t=1}^k \subset \{i\}_{i=1}^n\}$ is $2\delta M$ dense in H.

Proof: For any $x \in X$, let $B(x, \delta) = \{y \in X \mid d(x, y) \leq \delta\}$ be the closed δ ball. The union of the open balls $\{B^\circ(x, \delta) \mid x \in X\}$ covers X. By compactness, there exist $x_1, x_2, \ldots, x_n$ in X such that

$$X = \bigcup_{i=1}^n B^\circ(x_i, \delta) = \bigcup_{i=1}^n B(x_i, \delta).$$

So there is a finite set of closed sets which is δ-dense in all the closed sets. It is then routine to check that the test functions associated with these closed sets are $2M\delta$-dense in the set of all the test functions (associated with M). This completes the proof of the lemma.

Corollary. Let $\{h_{S_1}, \ldots, h_{S_m}\}$ be a $2M\delta$ dense subset of $H \subset M_\ell(C(X))$ in the lemma and let $\{x_i\}_{i=1}^n$ and $\{y_i\}_{i=1}^n$ be two groups of points on X. Suppose for $\varepsilon > 0$, the eigenvalues of $\{h_{S_j}(x_i)\}_{i=1}^n$ and $\{h_{S_j}(y_i)\}_{i=1}^n$ can be paired to within ε one by one, in increasing order, counting multiplicity, for $j = 1, 2, \ldots, m$; then $\{x_i\}_{i=1}^n$ and $\{y_i\}_{i=1}^n$ can be paired to within $3(\varepsilon + 4\delta M)$ one by one.

Proof: For any closed set $S \subset X$, there exists S_j such that $\|h_S - h_{S_j}\| < 2\delta M$. By the Weyl spectral variation inequality [1], the eigenvalues of $h_S(x_i)$ and $h_{S_j}(x_i)$ are within $2\delta M$ one by one. Since the eigenvalues of $\{h_{S_j}(x_i)\}_{i=1}^n$ and $\{h_{S_j}(y_i)\}_{i=1}^n$ can be paired to within ε one by one, the eigenvalues of $\{h_S(x_i)\}_{i=1}^n$ and $\{h_S(y_i)\}_{i=1}^n$ can be paired to within $\varepsilon + 4\delta M$ one by one. Notice that if we rearrange the pairing so that they are paired in increasing order the number $\varepsilon + 4\delta M$ will not be changed. Hence an application of Lemma 2.1 gives us the desired result.

2.4. Applying Lemma 2.3 together with the small eigenvalue variation theorem in [4], we prove a small spectrum variation theorem. One may think of this as a generalized small eigenvalue variation theorem. More precisely, consider a real rank zero C*-algebra $A = \lim_{\rightarrow}(A_n, \phi_{n,m})$ with $*$-homomorphisms $\phi_{n,m} : A_n \to A_m$. Assume that each A_n has the form

$$A_n = \bigoplus_{k=1}^{r_n} C(X_{(nk)}) \otimes M_{[n,k]}$$

where r_n is finite, $X_{(nk)}$ is a compact metric space, $[n,k]$ is a positive integer and $M_{[n,k]}$ is the C*-algebra of $[n,k] \times [n,k]$ matrices.

For $m > n$, fix a point w in $X_{(mk)}$, we have the following $*$-homomorphism

$$A_n \to A_m \xrightarrow{\pi_k} C\big(X_{(mk)}\big) \otimes M_{[m,k]} \xrightarrow{\tau_w} M_{[m,k]}$$

where π_k is the quotient map from A_m to its kth summand and τ_w is the evaluation map at w. (For convenience, we may assume that $\phi_{n,m}$ is unital. For the non-unital case, one needs only a small modification.) This gives rise to a representation $\phi_w : A_n \to M_{[m,k]}$. So there are r_n groups of points

$$\left(\{x_{1t}\}_{t=1}^{k_1}, \ldots, \{x_{r_n t}\}_{t=1}^{k_{r_n}}\right) \subset (X_{(n1)}, \ldots, X_{(nr_n)})$$

and a unitary $u_w \in M_{[m,k]}$ such that

$$\phi_w(f) = u_w \begin{bmatrix} f(x_{11}) & & & & & \\ & \ddots & & & & \\ & & f(x_{1k_1}) & & & \\ & & & \ddots & & \\ & & & & f(x_{r_n 1}) & \\ & & & & & \ddots \\ & & & & & & f(x_{r_n k_{r_n}}) \end{bmatrix} u_w^*$$

for all $f \in A_n$. Clearly, these points may change if w changes.

Theorem. Let $A = \lim_{\to}(A_n, \phi_{n,m})$ be as above. Fix n and fix an $\varepsilon > 0$ $(\varepsilon < 1)$, there exists $m > n$ such that for any two points in $X_{(mk)}$, $1 \leq k \leq r_m$, the corresponding two groups of points on $\bigcup_{k=1}^{r_n} X_{(nk)}$, can be paired, inside each of $\{X_{(nk)}\}_{k=1}^{r_n}$, to within ε one by one. Moreover, the result holds for all $j > m$.

Proof: For any m, any self-adjoint element $g \in A_m$, any k and any $w \in X_{(mk)}$, define $\lambda_t^{(k)}(w) = $ the tth lowest eigenvalue of $g(w)$, counting multiplicity. The eigenvalue variation $EV(g)$ was defined in [4] as follows:

$$EV(g) = \sup \left\{ |\lambda_t^{(k)}(w) - \lambda_t^{(k)}(w')| \,\Big|\, w, w' \in X_{(mk)}, \quad 1 \leq k \leq r_m, \quad \text{all possible} \quad t \right\}.$$

In [4] it was shown that for any finitely many self-adjoint elements $\{h_i\}_{i=1}^a \subset A_n$ and for $\varepsilon' > 0$, there exists $m > n$ such that

$$EV\big(\phi_{n,m}(h_i)\big) < \varepsilon' \qquad i = 1, 2, \ldots, a.$$

Furthermore, this holds for all $j > m$.

Let $M = \frac{1}{\varepsilon}$ and $\delta = \varepsilon^2$. We can form the test functions $H_k \subset C(X_{(nk)}) \otimes M_{[nk]}$ associated with M as in 2.3. By Lemma 2.4, there are finite subsets $\{\widetilde{H}_k\}_{k=1}^{r_n}$ which are $2\delta M = 2\varepsilon$ dense in $\{H_k\}_{k=1}^{r_n}$. Extend each test function in $C(X_{(nk)}) \otimes M_{[nk]}$ to be zero in other summands of A_n. Notice that the unit of each summand is in $\bigcup_{k=1}^{r_n} \widetilde{H}_k$.

Apply the eigenvalue variation theorem to $\bigoplus_{k=1}^{r_n} \widetilde{H}_k$ and $\varepsilon' = \frac{\varepsilon}{15}$; there exists $m > n$ such that $EV\big(\phi_{n,m}(h)\big) < \varepsilon'$ for all $h \in \bigoplus_{k=1}^{r_n} \widetilde{H}_k$. Furthermore, this holds for all $j > m$.

Let w and w' be two points in $X_{(mk)}$. There are two groups of points associated with them, say $\big(\{x_{1t}\}_{t=1}^{k_1}, \ldots, \{x_{r_n t}\}_{t=1}^{k_{r_n}}\big)$ and $\big(\{x'_{1t}\}_{t=1}^{k'_1}, \ldots, \{x'_{r_n t}\}_{t=1}^{k'_{r_n}}\big)$. Our first conclusion is that

$$(k_1, \ldots, k_{r_n}) = (k', \ldots, k'_{r_n}).$$

In fact, let h be the unit of $C(X_{(nt)}) \otimes M_{[n,t]}$; then $h \in \widetilde{H}_t$ and $EV\big(\phi_{n,m}(h)\big) < \varepsilon'$ implies $k_t = k'_t$.

For $h \in \widetilde{H}_t$, the eigenvalues of $\{h(x_{tj})\}_{j=1}^{k_t}$ and $\{h(x'_{tj})\}_{j=1}^{k_t}$ are the same as the eigenvalues of $\phi_w(h)$ and $\phi_{w'}(h)$, respectively. Since the eigenvalues of $\phi_w(h)$ and $\phi_{w'}(h)$ are within ε' one by one, $\{x_{tj}\}_{j=1}^{k_t}$ and $\{x'_{tj}\}_{j=1}^{k_t}$ can be paired to within $3(\varepsilon' + 4(\varepsilon')^2/\varepsilon') = 15\varepsilon' = \varepsilon$ one by one.

Since this is true for $t = 1, 2, \ldots, r_n$, we have proved the first part of the theorem. It is obvious that this holds for all $j > m$.

2.5. For the purpose of later use we translate the terminologies and results into the case that the compact metric spaces are finite connected graphs. We may suppose that a graph is metricized by its arclength with each edge of length one.

Fix $M \geq 1$ and a large positive integer N. Let $H \subseteq C(X) \otimes M_\ell$ be the test functions associated with M, where X is a finite connected graph and ℓ a positive integer. We can form a $2M/N$ dense finite subset as follows. Divide each edge of X into N small intervals. Let $\{S\}$ be those closed sets whose boundaries are among the partition points.

Lemma. Let $\{x_i\}_{i=1}^n$ and $\{y_i\}_{i=1}^n$ be two groups of points in a finite connected graph X. Then:

(i) $\{h_S\}$ is $2M/N$ dense in H.

(ii) If the eigenvalues of $\{h_S(x_i)\}_{i=1}^n$ and $\{h_S(y_i)\}_{i=1}^n$ are within a given $\varepsilon > 0$ one by one for all $S \in \{S\}$, then the points $\{x_i\}_{i=1}^n$ and $\{y_i\}_{i=1}^n$ can be paired to within $3(\varepsilon + 2M/N)$ one by one.

In some places, we may need to use the self-adjoint elements $P h_S P$ where P is a diagonal constant projection in $C(X) \otimes M_\ell$. We will include them among our test functions.

Chapter 3.   Perturbation

In this section we will prove a perturbation result Theorem 3.1 for certain $*$- homomorphisms.

3.1.   The maps we are interested in are those from finite direct sums of basic building blocks to finite direct sums of basic building blocks. By passing to quotients, we may reduce the second algebra to a single basic building block. We will only prove the case that the first algebra is also a single basic building block. The proof for finitely many ones will be the same. This will be clear from the proof. The special case that the graphs are Hausdorff intervals and circles is proved in [11].

Theorem.   Let A and B be two basic building blocks with spectra X and Y and with generic fibres M_n and M_m, respectively. For any unital $*$-homomorphism ϕ from A to B, any finite subset $F \subset A$ and any $\varepsilon > 0$, there exists a unital $*$-homomorphism ϕ' from A to B such that:

(1) $\|\phi(f) - \phi'(f)\| < \varepsilon$ for $f \in F$.

(2)   On each edge of Y, identified with $I = [0, 1]$,

$$\phi'(f)(t) = W(t) \begin{bmatrix} f\big(s_1(t)\big) & & \\ & \ddots & \\ & & f\big(s_p(t)\big) \end{bmatrix} W^*(t)$$

for all $f \in A$ and all $t \in I$, where $W(\cdot)$ is a unitary in $M_m\big(C(I)\big)$ and $\{s_i(\cdot)\}_{i=1}^p \subset C(I, X)$.

Proof: Fix $\delta > 0$, $\gamma > 0$, $\beta > 0$ and three integers $1 < R < M < N$ to be specified later.

For R and M, we can form the test functions associated with them in $C(X) \otimes M_n$. They are in fact in A. As in 2.5, partition each edge of X into N intervals and let H and $\widetilde{H}$ be the finitely many test functions that are $\frac{2M}{N}$ and $\frac{2R}{N}$ dense in all the test functions associated with M and R respectively.

We need another finite subset $\widetilde{F} \subset A$. The elements we are interested in are those coming from matrix units of M_n. Let $e_{kj} \in M_n$ be one of the canonical matrix units which is one at (k, j) position and zero at other positions. We may regard e_{kj} as an element in $C(X) \otimes M_n$.

However, e_{kj} may not belong to A due to the multiple point property at the vertices of X. We now construct an element $\widetilde{e_{kj}}$ in A. Let $S \subset X$ be the set of those vertices of X such that e_{kj} is not in their fibres. For each $s \in S$ let $B_s = \{x \in X \mid d(x, s) \leq \varepsilon\}$. Notice that we have finitely many disjoint ε balls $(\varepsilon < 1/2)$. Define

$$a_{kj}(x) = \begin{cases} 0 & x \in S \\ \text{linear with slope } \frac{1}{\varepsilon} & x \in \cup B_s \\ 1 & x \in X \backslash (\bigcup_{s \in S}) B_s \end{cases}$$

Clearly, $a_{kj} \in C(X)$. $\tilde{e}_{kj}$ is defined as $a_{kj}e_{kj}$. The set $\widetilde{F}$ will consist of all the $\tilde{e}_{kj}$ and all the possible orthogonal sums of $\{\tilde{e}_{jj} \mid j = 1, 2, \ldots, n\}$.

Partition each edge of Y into small intervals so that the variations of $\phi(h)$ and $\phi(\tilde{h})$ are within δ for $h \in H$ and $\tilde{h} \in \widetilde{H}$ and the variations of $\psi(f)$ and $\phi(\tilde{f})$ are within γ for $f \in F$ and $\tilde{f} \in \widetilde{F}$, on each small interval.

We are now ready to construct ϕ'. We will first define ϕ' at each vertex of Y and then interpolate it on each edge.

Let $y_0 = \{y_j^{(0)}\}_{j=1}^a$ be a vertex of Y with multiplicity a. This is to say that the fibre of B at y_0 has diagonal block form. Denote by τ_{y_0} the evaluation map of B at y_0 and by π_i the quotient map from the fibre of B at y_0 onto its ith block. Then the map $\pi_i \circ \tau_{y_0} \circ \phi$ is a $*$-homomorphism from A to the ith block of the fibre of B at y_0. Hence for $f \in A$,

$$\pi_i \circ \tau_{y_0} \circ \phi(f) = u_i \begin{bmatrix} f(s_1) & & \\ & \ddots & \\ & & f(s_p) \end{bmatrix} u_i^*$$

where $\{s_i\}_{i=1}^p \subset X$ and u_i is a unitary matrix in the ith block of the fibre of B at y_0.

Let us look at $\{s_i\}_{i=1}^p$. Some of the vertex points may form a single vertex. We may group these together and count them as one point in $\{s_i\}_{i=1}^p$. In other words, those vertex points listed can not form a full vertex. Move some of $\{s_i\}_{i=1}^p$ to nearby points so that all the points are distinct except those vertex points and so that there is no full vertex. Denote these new points by $\{\tilde{s}_i\}_{i=1}^p$. We shall say that the points $\{\tilde{s}_i\}_{i=1}^p$ have the distinctness property.

Define a representation ϕ_i' from A to the ith block of the fibre of B at y_0 as follows:

$$\phi_i'(f) = u_i \begin{bmatrix} f(\tilde{s}_1) & & \\ & \ddots & \\ & & f(\tilde{s}_p) \end{bmatrix} u_i^* \quad f \in A.$$

It is clear that if the movement is very small, we will have the following inequality:

$$\|\phi_i'(f) - \pi_i \circ \tau_{y_0} \circ \phi(f)\| < \varepsilon'$$

for $f \in F \cup \widetilde{F} \cup H \cup \widetilde{H}$ and for some $\varepsilon' < \varepsilon$ to be fixed later.

This construction can be carried out for $i = 1, 2, \ldots, a$. Finally, we have a representation ϕ' from A to the fibre of B at y_0 such that

$$\phi'(f) = u \begin{bmatrix} f(s_1) & & \\ & \ddots & \\ & & f(s_q) \end{bmatrix} u^*$$

for all $f \in A$ and such that

$$\|\phi'(y_0)(f) - \phi(f)(y_0)\| < \varepsilon'$$

for all $f \in F \cup \widetilde{F} \cup H \cup \widetilde{H}$.

In this way, we can define a $*$-homomorphism ϕ' from A to the fibres of B at the vertices of Y so that the difference between ϕ' and ϕ on $F \cup \widetilde{F} \cup H \cup \widetilde{H}$ is within ε'.

To extend ϕ' over Y, it is enough to do that on each edge of Y. So we identify an edge L of Y with $I = [0, 1]$ and we will interpolate ϕ' on $(0, 1)$. Recall that there was a partition on I, say $0 = t_0 < t_1 < t_2 < \cdots < t_K < t_{K+1} = 1$. We will start with $[t_1, t_2]$. The construction of ϕ' will be divided into four steps.

Step 1. Define ϕ' on $\left[t_1, \frac{t_1+t_2}{2}\right]$.

Choose representations of ϕ at t_1 and t_2, say

$$\phi(f)(t_1) = U \begin{bmatrix} f(\lambda_1) & & \\ & \ddots & \\ & & f(\lambda_p) \end{bmatrix} U^*$$

$$\phi(f)(t_2) = V \begin{bmatrix} f(\xi_1) & & \\ & \ddots & \\ & & f(\xi_q) \end{bmatrix} V^*$$

for all $f \in A$, where U and V are two unitaries in M_m and $\{\lambda_i\}_{i=1}^p \cup \{\xi_i\}_{i=1}^q \subset X$. Move these points around so that they have the distinctness property. We use the same notations to denote these new points. Replace those old points in the representations of ϕ at t_1 and t_2 by new points. We have then defined ϕ' at t_1 and t_2. We may make the moving so small that

$$\|\phi'(f)(t_1) - \phi'(f)(t_2)\| < \gamma \qquad f \in F \cup \widetilde{F}$$
$$\|\phi'(h)(t_1) - \phi'(h)(t_2)\| < \delta \qquad h \in H \cup \widetilde{H}.$$

still holds.

We would like to show that $p = q$, provided that $\delta < \frac{1}{8m}$, $\frac{8}{M} < \delta$ and $\frac{8M}{N} < \delta$.

Let $x = \{x_i\}_{i=1}^b$ be a vertex of X. We assert that the multiplicities of each x_i in $\{\lambda_i\}_{i=1}^p$ and in $\{\xi_i\}_{i=1}^q$ are the same. As a consequence, $p = q$.

To show this, let $L_1, L_2, \ldots$ and L_k be the edges of X joining at x and identify each of them with $I = [0, 1]$, with or without 0 and 1 being identified, where x is identified with 1. On each L_j, there must be an open interval S_j with length δ such that

$$\left(\bigcup_{j=1}^{k} S_j \right) \cap \left(\{\lambda_i\}_{i=1}^{p} \cup \{\xi_i\}_{i=1}^{q} \right) = \emptyset .$$

This is because that there are at most $p + q \leq 2m$ points but there are at least $\left[\frac{1}{2} \frac{1}{\delta} \right]$ disjoint open intervals with length δ inside $\left[0, \frac{1}{2} \right]$ and $\left[\frac{1}{2}, 1 \right]$, respectively, where $\left[\frac{1}{2} \frac{1}{\delta} \right]$ is the largest integer less than $\frac{1}{2\delta}$. Since $\delta < \frac{1}{8n}$, we have $\left[\frac{1}{2} \frac{1}{\delta} \right] \geq \left[\frac{1}{2} 8m \right] = 4m > 2m$. So we conclude that on each L_j, one of the length δ open intervals, say S_j, does not contain any of $\{\lambda_i\}_{i=1}^{p} \cup \{\xi_i\}_{i=1}^{q}$.

Recall that we had a N-partition on each edge of X. There are $\frac{1}{\left[\frac{1}{\delta} \right]} N - 2$ partition points in each S_j. So we do have points in each S_j if $\frac{6}{N} < \delta$. Take two partition points $\{c_j, d_j\}$ such that they are at least $\frac{\delta}{2} - \frac{2}{N} > \frac{4}{M} - \frac{2}{N} = \frac{2}{M}$ from the boundary of S_j. Let S be the union of all the circles in $\{L_j\}_{j=1}^{k}$ together with the union of $[d_j, 1] \subset L_j$ of those non-circle edges L_j. Then S is a neighbourhood of x and $h_S \in H$.

For $t \in \{\lambda_i\}_{i=1}^{p} \cup \{\xi_i\}_{i=1}^{q}$, $t \notin S_j$, $h_S(t)$ is either 0 or 1. This will ensure that the multiplicities of each of x_j in $\{\lambda_i\}_{i=1}^{p}$ and $\{\xi_i\}_{i=1}^{q}$ are the same. To see this, assume the contrary. Let $a_1 < a_2$ be the multiplicities of x_j in $\{\lambda_i\}_{i=1}^{p}$ and in $\{\xi_i\}_{i=1}^{q}$, respectively. Denote by $P \in C(X) \otimes M_n$ the diagonal constant projection corresponding to x_j. Then $Ph_S P \in H$, and we have

$$\| \phi'(Ph_S P)(t_1) - \phi'(Ph_S P)(t_2) \| < \delta < 1 .$$

This says that there must be $a_2 - a_1$ more points in $\{\xi_i\}_{i=1}^{q} \cap \{S \setminus \{x\}\}$ than in $\{\lambda_i\}_{i=1}^{p} \cap \{S \setminus \{x\}\}$. Now let $Q \in C(X) \otimes M_n$ be the diagonal constant projection corresponding to x_i; the similar argument says that there must be $a_2 - a_1$ more x_i in $\{\xi_j\}_{j=1}^{q}$ than in $\{\lambda_i\}_{i=1}^{p}$. Hence there are $a_2 - a_1$ more $x = \{x_i\}_{i=1}^{b}$ in $\{\xi_i\}_{i=1}^{q}$ than in $\{\lambda_i\}_{i=1}^{p}$. But this contradicts the assumption that the vertex points in $\{\xi_j\}_{j=1}^{q}$ can not form a whole vertex of X. So we must have $a_1 = a_2$. It is easy to see now that $p = q$.

As a byproduct, we have proved that $\{\lambda_i\}_{i=1}^{p}$ and $\{\xi_i\}_{i=1}^{p}$ can be paired to within $m\delta$ one by one. We would like to show that we can replace $m\delta$ by another number $3(2^a \delta + \frac{4M}{N})$, independent of m, where a is the total number of vertex points of X. We will use this result later. The way we are going to show what we have just stated is first to remove vertex points and second to apply the pairing lemma.

For a self-adjoint element $h \in C(X) \otimes M_n$, the eigenvalues of $\phi'(h)(t_1)$ are exactly the same as the eigenvalues of $\{h(\lambda_i)\}_{i=1}^{p}$ and the eigenvalues of $\phi'(h)(t_2)$ are the same as the eigenvalues of $\{h(\xi_i)\}_{i=1}^{p}$, counting multiplicities. So the condition

$$\| \phi'(h)(t_1) - \phi'(h)(t_2) \| < \delta$$

together with the Weyl spectral variation inequality [1] ensure that the eigenvalues of $\{h(\lambda_i)\}_{i=1}^p$ and the eigenvalues of $\{h(\xi_i)\}_{i=1}^p$ can be paired to within δ one by one, counting multiplicities, in decreasing order.

Let $x = \{x_i\}_{i=1}^b$ be a vertex of X and let x_i be a point in $\{\lambda_j\}_{j=1}^p \cap \{\xi_j\}_{j=1}^p$. There are three cases:

(i) Those eigenvalues of $h(x_i)$ in $\{h(\lambda_i)\}_{i=1}^p$ happen to pair with the same eigenvalues in $\{h(\xi_i)\}_{i=1}^p$ to within δ. After removing x_i, the eigenvalues of $\{h(\lambda_i)\}_{\lambda_i \neq x_i}$ and the eigenvalues of $\{h(\xi_i)\}_{\xi_i \neq x_i}$ are still within δ one by one

(ii) One of the eigenvalues of $h(x_i)$, say λ, pairs with an eigenvalue of $\{h(\xi_i)\}_{i=1}^p$, say d, and assume $\lambda \leq d$. We may assume that d is the largest one with this property. Noticing that λ is also an eigenvalue of $\{h(\xi_i)\}_{i=1}^p$. So λ pairs with an eigenvalue of $\{h(\lambda_i)\}_{i=1}^p$. We denote the smallest one by η. Hence $0 \leq d - \lambda \leq \delta$ and $0 \leq \lambda - \eta \leq \delta$. If we remove x_i, λ disappears. Now d must pair with something no less than η and η must pair with something less or equal to d. The pairings of the eigenvalues outside $[\eta, d]$ are unchanged. Noticing that $d - \eta \leq (d - \lambda) + (\lambda - \eta) \leq 2\delta$ we conclude that the eigenvalues of $\{h(\lambda_j) \mid \lambda_j \neq x_i\}$ and the eigenvalues of $\{h(\xi_j) \mid \xi_j \neq x_i\}$ can be paired to within 2δ one by one.

(iii) $\lambda > d$. Similar to (ii), the eigenvalues of $\{h(\lambda_j) \mid \lambda_j \neq x_i\}$ and the eigenvalues of $\{h(\xi_j) \mid \xi_j \neq x_i\}$ can be paired to within 2δ one by one.

In any case, after removing x_i, the eigenvalues can be paired to within 2δ one by one. If we remove all the vertex points, at most a of them, the eigenvalues of $\{h(\lambda_i) \mid \lambda_i$ is not a vertex point$\}$ and the eigenvalues of $\{h(\xi_i) \mid \xi_i$ is not a vertex point$\}$ can be paired to within $2^a\delta$ one by one.

Now we are ready to apply the pairing lemma. Those remaining points can be paired to within $3(2^a\delta + \frac{2M}{N})$ one by one. Since the vertex points in $\{\lambda_i\}_{i=1}^p$ can be paired with the vertex points in $\{\xi_i\}_{i=1}^p$, we conclude that $\{\lambda_i\}_{i=1}^p$ and $\{\xi_i\}_{i=1}^p$ can be paired to within $3(2^a\delta + \frac{2M}{N})$ one by one.

Changing V by a permutation we may assume that the points in each pair appear in the same diagonal position of the representations of ϕ at t_1 and t_2. Define p continuous mappings from $\left[t_1, \frac{t_1+t_2}{2}\right]$ to X by moving $\{\lambda_i\}_{i=1}^p$ to $\{\xi_i\}_{i=1}^p$, within each pair, in a linear fashion. Denote these maps by $\{\lambda_i(t)\}_{i=1}^p$. Notice that for those small size blocks, the maps are constant.

Define ϕ' on $\left[t_1, \frac{t_1+t_2}{2}\right]$ as follows: for $f \in A$,

$$\phi'(f)(t) = U \begin{bmatrix} f(\lambda_1(t)) & & \\ & \ddots & \\ & & f(\lambda_p(t)) \end{bmatrix} U^*.$$

Clearly, for any $s, t \in \left[t_1, \frac{t_1 + t_2}{2}\right]$, we have

$$\|\phi'(f)(t) - \phi'(f)(s)\|$$
$$\leq \max_i \left\{\|f(\lambda_i(t)) - f(\lambda_i(s))\|\right\}.$$

Since $d(\lambda_i(t), \lambda_i(s)) < 3(2^a \delta + \frac{2M}{N})$, the right hand side of the above expression can be less than γ for $f \in F \cup \widetilde{F}$, if $\delta \ll \gamma$ and $\frac{M}{N} \ll \gamma$. For $\tilde{h} \in \widetilde{H}$,

$$\|\tilde{h}(\lambda_i(t)) - \tilde{h}(\lambda_i(s))\| < R \cdot \frac{1}{R^2} = \frac{1}{R}$$

if $3(2^a \delta + \frac{M}{N}) < \frac{1}{R^2}$. Here for fixed R we need $\frac{8}{M} < \delta$, $\frac{8M}{N} < \delta < 1$ and $3(2^a \delta + \frac{2M}{N}) < \frac{1}{R^2}$.

Step 2.   Interpolate ϕ' on the right hand half-interval of $\left[\frac{t_1 + t_2}{2}, t_2\right]$.

For $\beta > 0$ fixed, divide $\{\xi_i\}_{i=1}^p$ into groups. Within each group, each point is strictly less than distance 4β from some other point. Different groups are separated by at least 4β. Clearly, the diameter of each group is strictly less than $4m\beta$. Here, for the vertex points, we regard them as the vertices they are in. In other words, two different vertex points in a vertex are in the same group.

In each group, a point can be connected to another point in the same group via a minimal curve on X. The union of all those curves form a closed set. The boundary points of this set are in this group. Clearly, different groups correspond to different closed sets with distance at least 2β.

For a group, denote the closed set by S. We can construct a function, as in 2.6, with slope R, say $\tilde{h}_S$. Now $\tilde{h}_S = 1$ on this group of points and $\tilde{h}_S = 0$ on all other groups of points, if $\frac{1}{R} < \frac{\beta}{2}$. If N is large enough so that $\frac{1}{N} < \frac{\beta}{4}$, then we can extend S a little to a large set $\widetilde{S}$ such that $\tilde{h}_{\widetilde{S}} \in \widetilde{H}$. Such different extended closed sets are $2\beta - 2\frac{\beta}{4} = \frac{3}{2}\beta$ apart. So $\tilde{h}_{\widetilde{S}} = 1$ on this group of points and $\tilde{h}_{\widetilde{S}} = 0$ on all the other groups of points. Here we have two projections $\phi'(\tilde{h}_{\widetilde{S}})\left(\frac{t_1 + t_2}{2}\right)$ and $\phi'(\tilde{h}_{\widetilde{S}})(t_2)$. They are related in the following ways:

$$\phi'(\tilde{h}_{\widetilde{S}})\left(\frac{t_1 + t_2}{2}\right) = U \begin{bmatrix} \tilde{h}_{\widetilde{S}}(\xi_1) & & \\ & \ddots & \\ & & \tilde{h}_{\widetilde{S}}(\xi_p) \end{bmatrix} U^*,$$

$$\phi'(\tilde{h}_{\widetilde{S}})(t_2) = V \begin{bmatrix} \tilde{h}_{\widetilde{S}}(\xi_1) & & \\ & \ddots & \\ & & \tilde{h}_{\widetilde{S}}(\xi_p) \end{bmatrix} V^*.$$

We denote the projection $\mathrm{diag}(\tilde{h}_{\widetilde{S}}(\xi_1), \ldots, \tilde{h}_{\widetilde{S}}(\xi_p))$ by P.

Carrying out this process for each group of points we get projections, say $P_1, P_2, \ldots, P_r$. For convenience, we may write $P_1 = \begin{bmatrix} 1 & & & \\ & 0 & & \\ & & \ddots & \\ & & & 0 \end{bmatrix}, \ldots, P_r = \begin{bmatrix} 0 & & & \\ & \ddots & & \\ & & 0 & \\ & & & 1 \end{bmatrix}$, where each 1 is of suitable size.

For any $\tilde{h} \in \widetilde{H}$, we have the following inequality:

$$\left\| \phi'(\tilde{h}) \left(\frac{t_1 + t_2}{2} \right) - \phi'(\tilde{h})(t_2) \right\|$$

$$\leq \left\| \phi'(\tilde{h}) \left(\frac{t_1 + t_2}{2} \right) - \phi'(\tilde{h})(t_1) \right\| + \left\| \phi'(\tilde{h})(t_1) - \phi'(\tilde{h})(t_2) \right\|$$

$$\leq \frac{1}{R} + \delta < \frac{2}{R} \, .$$

In particular, we have

$$\| U P_i U^* - V P_i V^* \| < \frac{2}{R} \qquad i = 1, 2, \ldots, p$$

or

$$\| V^* U P_i - P_i V^* U \| < \frac{2}{R} \qquad i = 1, 2, \ldots, p.$$

Let us write the unitary $V^* U = \begin{bmatrix} w_{11} & w_{12} \\ w_{21} & w_{22} \end{bmatrix}$, where the size of w_{11} is the same as the rank of P_1. We then have $\| w_{12} \| < \frac{2}{R}$ and $\| w_{21} \| < \frac{2}{R}$. Applying this computation to all $P_1, P_2, \ldots, P_r$ we have

$$\left\| V^* U - \begin{bmatrix} w_1 & & \\ & \ddots & \\ & & w_r \end{bmatrix} \right\| < \frac{2}{R} r^2 \leq \frac{2}{R} m^2 \, .$$

Write $T = \begin{bmatrix} w_1 & & \\ & \ddots & \\ & & w_r \end{bmatrix}$. T is invertible if $\frac{2}{R} m^2 < 1$. There is a unitary $\widetilde{W}$ such that $T = |T^*| \widetilde{W}$. So

$$\| V^* U \widetilde{W}^* - |T^*| \, \| < \frac{2m^2}{R} \, .$$

Since $V^* U \widetilde{W}^*$ is a unitary and $|T^*|$ is positive, the eigenvalues of $|T^*|$ are close to 1 to within $\frac{2m^2}{R}$. So,

$$\| \, |T^*| - 1 \, \| < \frac{2m^2}{R} \, .$$

Hence we have

$$\| V^* U \widetilde{W}^* - 1 \| \leq \| V^* U \widetilde{W}^* - |T^*| \| + \| |T^*| - 1 \|$$

$$< \frac{2m^2}{R} + \frac{3m^2}{R} = \frac{5m^2}{R} \, .$$

Set $V^*U\widetilde{W}^* = W'$. Let $W'(t)$ be a unitary path in a $\frac{5m^2}{R}$ neighbourhood of I, defined on $\left[\frac{t_1+3t_2}{4}, t_2\right]$, such that $W'(t_2) = 1$ and $W'\left(\frac{t_1+3t_2}{4}\right) = W'$. Define ϕ' on $\left(\frac{t_1+3t_2}{4}, t_2\right)$ for all $f \in A$ as follows:

$$\phi'(f)(t) = VW'(t) \begin{bmatrix} f(\xi_1) & & \\ & \ddots & \\ & & f(\xi_p) \end{bmatrix} W'^*(t)V^*.$$

ϕ' is certainly continuous on $\left[\frac{t_1+3t_2}{4}, t_2\right]$. At t_2, we have

$$\phi'(f)(t_2) = V \begin{bmatrix} f(\xi_1) & & \\ & \ddots & \\ & & f(\xi_p) \end{bmatrix} V^*$$

which agrees with the previous definition. At the other end,

$$\phi'(f)\left(\frac{t_1 + 3t_2}{4}\right) = U\widetilde{W}^* \begin{bmatrix} f(\xi_1) & & \\ & \ddots & \\ & & f(\xi_p) \end{bmatrix} \widetilde{W}U^*.$$

Since $W'(t)$ is in a $\frac{5m^2}{R}$ neighbourhood of I, for $f \in A$, $t, s \in \left[\frac{t_1+3t_2}{4}, t_2\right]$,

$$\|\phi'(f)(t) - \phi'(f)(s)\| \le 4\frac{5m^2}{R}.$$

Hence the variation of ϕ' in $\left[\frac{t_1+3t_2}{4}, t_2\right]$, on the unit ball of A, is within $\frac{20m^2}{R}$.

Step 3.   Interpolate on $\left[\frac{t_1+t_2}{2}, \frac{t_1+3t_2}{4}\right] = [c, d]$.

For convenience, we may assume at the beginning that the elements of $F \cup \widetilde{F}$ have norm less than or equal to one. For $f \in F \cup \widetilde{F}$, we have the inequality

$$\|\phi'(f)(c) - \phi'(f)(d)\| < 3\gamma$$

if $\frac{20m^2}{R} < \gamma$. In matrix form, we have

$$\left\| U\widetilde{W}^* \begin{bmatrix} f(\xi_1) & & \\ & \ddots & \\ & & f(\xi_p) \end{bmatrix} \widetilde{W}U^* - U \begin{bmatrix} f(\xi_1) & & \\ & \ddots & \\ & & f(\xi_p) \end{bmatrix} U^* \right\| < 3\gamma.$$

Recall that the unitary $\widetilde{W}$ has diagonal form $\widetilde{W} = \operatorname{diag}(\widetilde{W}_{11}, \ldots, \widetilde{W}_{rr})$. Each block corresponds to a group of points. The above inequality becomes

$$
\left\| \begin{bmatrix} \widetilde{W}_{11}^* \begin{bmatrix} f(\xi_1) & & \\ & \ddots & \\ & & f(\xi_\ell) \end{bmatrix} & & \\ & \ddots & \\ & & \widetilde{W}_{rr}^* \begin{bmatrix} \ddots & \\ & f(\xi_p) \end{bmatrix} \end{bmatrix} \right.
$$
$$
\left. - \begin{bmatrix} \begin{bmatrix} f(\xi) & & \\ & \ddots & \\ & & f(\xi_\ell) \end{bmatrix} \widetilde{W}_{11}^* & & \\ & \ddots & \\ & & \begin{bmatrix} \ddots & \\ & f(\xi_p) \end{bmatrix} \widetilde{W}_{rr}^* \end{bmatrix} \right\| < 3\gamma \ .
$$

Here $\{\xi_i\}_{i=1}^{\ell}$ is assumed to be a group of points separated by 4β from the other groups.

Let us study the inequality

$$
\left\| \widetilde{W}_{11}^* \begin{bmatrix} f(\xi_1) & & \\ & \ddots & \\ & & f(\xi_\ell) \end{bmatrix} - \begin{bmatrix} f(\xi_1) & & \\ & \ddots & \\ & & f(\xi_\ell) \end{bmatrix} \widetilde{W}_{11}^* \right\| < 3\gamma
$$

in detail. Consider the following two cases:

Case 1.   The group is at least 2β away from any vertex.

In this case, $\tilde{e}_{kj}(\xi_t) \in M_m$ for $t = 1, 2, \ldots, \ell$ and for $k, j = 1, 2, \ldots, n$, with non-zero entries no less than $\frac{2\beta}{\varepsilon}$. Write $\widetilde{W}_{11}^* = (w_{ij})$, we have

$$
\| w_{st} \tilde{e}_{kj}(\xi_i) - \tilde{e}_{kj}(\xi_\ell) w_{st} \| < 3\gamma
$$

for $i, s, t = 1, 2, \ldots, \ell$ and for $j, k = 1, 2, \ldots, n$.

As for the matrix w_{st}, it commutes with the matrix units to within 3γ, so there exists $d_{st} \in \mathbb{C}$ such that

$$
\| w_{st} - d_{st} I_{st} \| < 3n^2 \gamma \frac{\varepsilon}{\beta} < 3m^2 \gamma \frac{\varepsilon}{\beta} \ .
$$

Here I_{st} is the identity matrix with suitable size. Let $D = (d_{st} I_{st})$; then

$$
\| \widetilde{W}_{11}^* - D \| < 3m^2 \ell^2 \gamma \frac{\varepsilon}{\beta} \ .
$$

Take $\beta \ll \varepsilon'$ so that $\|f(s) - f(t)\| < \varepsilon'$ whenever $d(s,t) < 2m\beta$ for all f in $F \cup \widetilde{F}$. Hence, for such an f,

$$\left\| D \begin{bmatrix} f(\xi_1) & & \\ & \ddots & \\ & & f(\xi_\ell) \end{bmatrix} - \begin{bmatrix} f(\xi_1) & & \\ & \ddots & \\ & & f(\xi_\ell) \end{bmatrix} D \right\|$$

$$\leq 2\|D\|\varepsilon' \leq 2\left(1 + 3m^2\ell^2\frac{\gamma\varepsilon}{\beta}\right)\varepsilon' .$$

Here we use the fact that D commutes with $\begin{bmatrix} f(\xi_1) & & \\ & \ddots & \\ & & f(\xi_1) \end{bmatrix}$.

Decompose $D = |D^*|O$ as in Step 2 in the commutant of the C*-algebra

$$\left\{ \begin{bmatrix} g(\xi_1) & & \\ & \ddots & \\ & & g(\xi_1) \end{bmatrix} \;\middle|\; g \in A_1 \right\} \cong M_n .$$

O commutes with $\begin{bmatrix} f(\xi_1) & & \\ & \ddots & \\ & & f(\xi_\ell) \end{bmatrix}$ to within $2\left(1+3m^2\ell^2\gamma\frac{\varepsilon}{\beta}\right)\varepsilon'$ for all f in $F \cup \widetilde{F}$. Connect O^* to I (with suitable size) by an exponential unitary path in that commutant, call it $O(t)$, on $\left[\frac{c+d}{2}, d\right]$. Then

$$\left\| \left[O(t), \begin{bmatrix} f(\xi_1) & & \\ & \ddots & \\ & & f(\xi_\ell) \end{bmatrix} \right] \right\| < 2\left(1 + 3m^2\ell^2\gamma\frac{\varepsilon}{\beta}\right)\varepsilon', \qquad f \in F \cup \widetilde{F} .$$

To define a unitary path on $\left[c, \frac{c+d}{2}\right]$, notice that

$$\|\widetilde{W}_{11}^* O^* - |D^*|\| < 3m^2\ell^2\gamma\frac{\varepsilon}{\beta}.$$

Applying the result in Step 2, we have

$$\|\widetilde{W}_{11}^* O^* - 1\| < 6m^2\ell^2\gamma\frac{\varepsilon}{\beta} .$$

Hence, there is a unitary path $X(t)$ on $\left[c, \frac{c+d}{2}\right]$ such that $X(c) = I$ and $X\left(\frac{c+d}{2}\right) = \widetilde{W}_{11}^* O^*$. Furthermore, the variation of $X(t)$ on $\left[c, \frac{c+d}{2}\right]$ is less than $12m^2\ell^2\gamma\frac{\varepsilon}{\beta}$.

Now for any $f \in A$ we have

$$\left\| X(t) \begin{bmatrix} f(\xi_1) & & \\ & \ddots & \\ & & f(\xi_\ell) \end{bmatrix} X^*(t) - X(s) \begin{bmatrix} f(\xi_1) & & \\ & \ddots & \\ & & f(\xi_\ell) \end{bmatrix} X^*(s) \right\|$$

$$\leq 36m^2\ell^2\frac{\gamma}{\beta}\varepsilon\|f\|$$

for t and s in $\left[c, \frac{c+d}{2}\right]$.

Define a unitary path $\widetilde{W}_{11}(t)$ on $[c, d]$ as follows:

$$
\widetilde{W}_{11}(t) = \begin{cases} X(t) & t \in \left[c, \frac{c+d}{2}\right] \\ \widetilde{W}_{11}^* O(t) & t \in \left[\frac{c+d}{2}, d\right]. \end{cases}
$$

Then $\widetilde{W}_{11}(c) = I$, $\widetilde{W}_{11}(d) = \widetilde{W}_{11}^*$ and $\widetilde{W}_{11}\left(\frac{c+d}{2}\right) = \widetilde{W}_{11}^* O^*$ is well-defined.

For any $f \in F \cup \widetilde{F}$ and $t, s \in [c, d]$, we have

$$
\left\| \widetilde{W}_{11}(t) \begin{bmatrix} f(\xi_1) & & \\ & \ddots & \\ & & f(\xi_\ell) \end{bmatrix} \widetilde{W}_{11}(t) - \widetilde{W}_{11}(s) \begin{bmatrix} f(\xi_1) & & \\ & \ddots & \\ & & f(\xi_\ell) \end{bmatrix} \widetilde{W}_{11}(s) \right\|
$$

$$
\leq \max\left(2(1 + 3m^2\ell^2 \frac{\gamma}{\beta})\varepsilon', \quad 36m^2\ell^2\frac{\gamma}{\beta}\varepsilon \right)
$$

provided $\|f\| \leq 1$. In other words, the variation for

$$
\widetilde{W}_{11}(t) \begin{bmatrix} f(\xi_1) & & \\ & \ddots & \\ & & f(\xi_\ell) \end{bmatrix} \widetilde{W}_{11}(t)
$$

on $[c, d]$ is within the above number.

Case 2. The group has a point within 2β with a vertex.

In this case, each point in this group, except vertex points, is at most $2m\beta$ away from that vertex, say $x = \{x_i\}_{i=1}^b$. Replace these ξ_i by $x = \{x_j\}_{j=1}^b$ and denote the corresponding matrix by $\begin{bmatrix} f(x_i) & & \\ & \ddots & \\ & & \end{bmatrix}$.

If $\beta \ll \varepsilon'$ and $\beta < \frac{\varepsilon\varepsilon'}{2m}$ then for $f \in F \cup \widetilde{F}$, we have

$$
\left\| \begin{bmatrix} f(\xi_1) & & \\ & \ddots & \\ & & f(\xi_\ell) \end{bmatrix} - \begin{bmatrix} f(x_i) & & \\ & \ddots & \\ & & \end{bmatrix} \right\| < \varepsilon'.
$$

This give us the following inequality:

$$\left\| \widetilde{W}_{11}^* \begin{bmatrix} f(x_i) & \\ & \ddots \end{bmatrix} \widetilde{W}_{11} - \begin{bmatrix} f(x_i) & \\ & \ddots \end{bmatrix} \right\|$$

$$\leq \left\| \widetilde{W}_{11}^* \left(\begin{bmatrix} f(x_i) & \\ & \ddots \end{bmatrix} - \begin{bmatrix} f(\xi_1) & & \\ & \ddots & \\ & & f(\xi_\ell) \end{bmatrix} \right) \right\|$$

$$+ \left\| \widetilde{W}_{11}^* \begin{bmatrix} f(\xi_1) & & \\ & \ddots & \\ & & f(\xi_\ell) \end{bmatrix} - \begin{bmatrix} f(\xi_1) & & \\ & \ddots & \\ & & f(\xi_\ell) \end{bmatrix} \widetilde{W}_{11}^* \right\|$$

$$+ \left\| \left(\begin{bmatrix} f(\xi_1) & & \\ & \ddots & \\ & & f(\xi_\ell) \end{bmatrix} - \begin{bmatrix} f(x_i) & \\ & \ddots \end{bmatrix} \right) \widetilde{W}_{11}^* \right\|$$

$$< \varepsilon' + 3\gamma + \varepsilon' < 5\varepsilon'$$

for $f \in F \cup \widetilde{F}$.

Let Y be a permutation matrix such that

$$Y \begin{bmatrix} \begin{bmatrix} f(x_1) & & \\ & \ddots & \\ & & f(x_1) \end{bmatrix} & & \\ & \ddots & \\ & & \begin{bmatrix} f(x_b) & & \\ & \ddots & \\ & & f(x_b) \end{bmatrix} \end{bmatrix} Y^* = \begin{bmatrix} f(x_i) & \\ & \ddots \end{bmatrix}.$$

Then

$$\left\| Y^* \widetilde{W}_{11}^* Y \begin{bmatrix} f(x_1) & & \\ & \ddots & \\ & & f(x_b) \end{bmatrix} Y^* \widetilde{W}_{11} Y - \begin{bmatrix} f(x_1) & & \\ & \ddots & \\ & & f(x_b) \end{bmatrix} \right\| < 5\varepsilon'$$

for $f \in F \cup \widetilde{F}$, $\|f\| \leq 1$.

We now approximate $Y^* \widetilde{W}_{11}^* Y$ by a diagonal matrix. First, write $Y^* \widetilde{W}_{11}^* Y$ as block form

$$Y^* \widetilde{W}_{11}^* Y = \begin{bmatrix} V_{11} & \cdots & V_{ib} \\ & \cdots & \\ V_{b1} & \cdots & V_{bb} \end{bmatrix}$$

according to $\{x_i\}_{i=1}^b$. Take $f \in \widetilde{F}$, for example, to be the unit for the block corresponding to x_1; then

$$\left\| \begin{bmatrix} V_{11} & \cdots & V_{ib} \\ 0 & \cdots & 0 \\ & \cdots & \\ 0 & \cdots & 0 \end{bmatrix} - \begin{bmatrix} V_{11} & 0 & 0 \\ V_{21} & \vdots & \vdots & \vdots \\ \vdots & \vdots & \vdots & \vdots \\ V_{b1} & 0 & 0 \end{bmatrix} \right\| < 5\varepsilon'.$$

This implies, if we go through all $x_1, \ldots, x_b$, that

$$\left\| Y^* \widetilde{W}_{11}^* Y - \begin{bmatrix} V_{11} & & \\ & \ddots & \\ & & V_{bb} \end{bmatrix} \right\| < 10b\varepsilon'.$$

As $\mathrm{diag}(V_{11}, \ldots, V_{bb})$ has polar decomposition

$$\begin{bmatrix} V_{11} & & \\ & \ddots & \\ & & V_{bb} \end{bmatrix} = \begin{bmatrix} |V_{11}| & & \\ & \ddots & \\ & & |V_{bb}| \end{bmatrix} \begin{bmatrix} \widetilde{V}_{11} & & \\ & \ddots & \\ & & \widetilde{V}_{bb} \end{bmatrix}$$

then

$$\left\| Y^* \widetilde{W}_{11}^* Y - \begin{bmatrix} \widetilde{V}_{11} & & \\ & \ddots & \\ & & \widetilde{V}_{bb} \end{bmatrix} \right\| < 20b\varepsilon'$$

as in Step 2. Furthermore, for $f \in F \cup \widetilde{F}$,

$$\left\| \begin{bmatrix} \widetilde{V}_{11} & & \\ & \ddots & \\ & & \widetilde{V}_{bb} \end{bmatrix} \begin{bmatrix} f(x_1) & & \\ & \ddots & \\ & & f(x_b) \end{bmatrix} - \begin{bmatrix} f(x_1) & & \\ & \ddots & \\ & & f(x_b) \end{bmatrix} \begin{bmatrix} \widetilde{V}_{11} & & \\ & \ddots & \\ & & \widetilde{V}_{bb} \end{bmatrix} \right\|$$
$$< 20b\varepsilon' + 5\varepsilon' + 20b\varepsilon' < (40b + 5)\varepsilon'.$$

Just as in Case 1, after trying all those possible matrix units in $\widetilde{F}$, we conclude that for each $\widetilde{V}_{ii}$, there exists a unitary O_{ii}, commuting with the block $\begin{bmatrix} f(x_i) & & \\ & \ddots & \\ & & f(x_i) \end{bmatrix}$ for all $f \in A$, such that

$$\| \widetilde{V}_{ii} - O_{ii} \| < L_i(40b + 5)\varepsilon'$$

where $L_i = 2 \ (\text{multiplicity of } x_i \text{ in } \{\xi_j\}_{j=1}^\ell)^2 \ (\text{size of } f(x_i))^2$. Hence, we have the following inequalities:

$$\left\| Y^* \widetilde{W}_{11}^* Y - \begin{bmatrix} O_{11} & & \\ & \ddots & \\ & & O_{bb} \end{bmatrix} \right\| < L(40b + 5)\varepsilon' + 20b\varepsilon' + 20b\varepsilon'$$

or

$$\left\| \widetilde{W}_{11}^* Y \begin{bmatrix} O_{11} & & \\ & \ddots & \\ & & O_{bb} \end{bmatrix}^* Y^* - 1 \right\| < L(40b + 5)\varepsilon' + 20b\varepsilon' + 20b\varepsilon'$$

where $L = \max L_i < 2a^2n^2$.

Let $X(t)$ be a unitary path on $\left[c, \frac{c+d}{2}\right]$ such that $X(c) = 1$ and $X\left(\frac{c+d}{2}\right) = \widetilde{W}_{11}^* YO^*Y^*$, where $O = \mathrm{diag}(O_{11}, \ldots, O_{bb})$. The variation of $X(t)$ on $\left[\frac{c+d}{2}\right]$ is less than $2\left(L(40b+5)\varepsilon' + 20b\varepsilon'\right) \le 260ba^3n^2\varepsilon'$.

Let $O(t)$ be a unitary path in the commutant of

$$\left\{ \begin{bmatrix} f(x_1) & & \\ & \ddots & \\ & & f(x_b) \end{bmatrix} \ \Bigg| \ f \in A \right\}$$

defined on $\left[\frac{c+d}{2}, d\right]$ such that $O\left(\frac{c+d}{2}\right) = O^*$ and $O(d) = I$. Then for $s, t \in \left[\frac{c+d}{2}, d\right]$ and $f \in F \cup \widetilde{F}$ we have the following estimate:

$$\left\| \widetilde{W}_{11}^* YO(t)Y^* \begin{bmatrix} f(\xi_1) & & \\ & \ddots & \\ & & f(\xi_\ell) \end{bmatrix} YO^*(t)Y^*\widetilde{W}_{11} \right.$$

$$\left. - \widetilde{W}_{11}^* YO(s)Y^* \begin{bmatrix} f(\xi_1) & & \\ & \ddots & \\ & & f(\xi_\ell) \end{bmatrix} YO^*(s)Y^*\widetilde{W}_{11} \right\|$$

$$= \left\| O(t)Y^* \begin{bmatrix} f(\xi_1) & & \\ & \ddots & \\ & & f(\xi_\ell) \end{bmatrix} YO^*(t) - O(s)Y^* \begin{bmatrix} f(\xi_1) & & \\ & \ddots & \\ & & f(\xi_\ell) \end{bmatrix} YO^*(s) \right\|$$

$$\le \left\| O(t)Y^* \begin{bmatrix} f(x_i) & & \\ & \ddots & \\ & & \end{bmatrix} YO^*(t) - O(s)Y^* \begin{bmatrix} f(x_i) & & \\ & \ddots & \\ & & \end{bmatrix} YO^*(s) \right\|$$

$$+ \left\| O(t)Y^* \left(\begin{bmatrix} f(\xi_1) & & \\ & \ddots & \\ & & f(\xi_\ell) \end{bmatrix} - \begin{bmatrix} f(x_i) & & \\ & \ddots & \\ & & \end{bmatrix} \right) YO^*(t) \right\|$$

$$+ \left\| O(s)Y^* \left(\begin{bmatrix} f(\xi_1) & & \\ & \ddots & \\ & & f(\xi_\ell) \end{bmatrix} - \begin{bmatrix} f(x_i) & & \\ & \ddots & \\ & & \end{bmatrix} \right) YO^*(s) \right\|$$

$$\le \left\| O(t) \begin{bmatrix} f(x_1) & & \\ & \ddots & \\ & & f(x_b) \end{bmatrix} O^*(t) - O(s) \begin{bmatrix} f(x_1) & & \\ & \ddots & \\ & & f(x_b) \end{bmatrix} O^*(s) \right\| + 2\varepsilon'$$

$$= 2\varepsilon' .$$

Define a unitary path on $[c, d]$ as follows:

$$\widetilde{W}_{11}(t) = \begin{cases} X(t) & t \in \left[c, \frac{c+d}{2}\right] \\ \widetilde{W}_{11}^* YO(t)Y^* & t \in \left[\frac{c+d}{2}, d\right] . \end{cases}$$

Then $\widetilde{W}_{11}(c) = I$, $\widetilde{W}_{11}(d) = \widetilde{W}_{11}^*$. The variation of

$$\widetilde{W}_{11}(t) \begin{bmatrix} f(\xi_1) & & \\ & \ddots & \\ & & f(\xi_\ell) \end{bmatrix} \widetilde{W}^*(t)$$

is less than $260 n^2 a^3 \varepsilon'$ on $[c, d]$.

Similarly, we may construct unitary paths for other groups of points, say $\widetilde{W}_{22}(t)$, ..., $\widetilde{W}_{rr}(t)$.
Let

$$\widetilde{W}(t) = \begin{bmatrix} \widetilde{W}_{11}(t) & & \\ & \ddots & \\ & & \widetilde{W}_{rr}(t) \end{bmatrix}.$$

Then for $t, s \in [c, d]$, we have

$$\| \widetilde{W}(t) \begin{bmatrix} f(\xi_1) & & \\ & \ddots & \\ & & f(\xi_p) \end{bmatrix} \widetilde{W}^*(t) - \widetilde{W}(s) \begin{bmatrix} f(\xi_1) & & \\ & \ddots & \\ & & f(\xi_p) \end{bmatrix} \widetilde{W}^*(s) \|$$
$$< 260 n^2 a^3 \varepsilon'$$

for $f \in F$.

Define ϕ' on $[c, d]$ as follows:

$$\phi'(f)(t) = U \widetilde{W}(t) \begin{bmatrix} f(\xi_1) & & \\ & \ddots & \\ & & f(\xi_p) \end{bmatrix} \widetilde{W}^*(t) U^*$$

where $t \in [c, d]$ and $f \in A$. The variation of $\phi'(f)(t)$ on $[c, d]$ is less than $260 \, n^2 a^3 \varepsilon'$ for $f \in F$.

Finally, define a unitary on $[t_1, t_2]$ as follows:

$$W(t) = \begin{cases} U & t \in \left[t_1, \frac{t_1+t_2}{2}\right] \\ U\widetilde{W}(t) & t \in \left[\frac{t_1+t_2}{2}, \frac{t_1+3t_2}{4}\right] \\ VW'(t) & t \in \left[\frac{t_1+3t_2}{2}, t_2\right] \end{cases}$$

and define p maps:

$$s_i(t) = \begin{cases} \lambda_i(t) & t \in \left[t_1, \frac{t_1+t_2}{2}\right] \\ \xi_i & t \in \left[\frac{t_1+t_2}{2}, t_2\right] \end{cases} \quad i = 1, 2, \ldots, p.$$

Using these notations, we have

$$\phi'(f)(t) = W(t) \begin{bmatrix} f(s_1(t)) & & \\ & \ddots & \\ & & f(s_p(t)) \end{bmatrix} W^*(t)$$

for $t \in [t_1, t_2]$ and $f \in A$.

For $f \in F$, the variation of $\phi'(f)$ on $\left[t_1, \frac{t_1+t_2}{2}\right]$ is within γ, the variation on $\left[\frac{t_1+t_2}{2}, \frac{t_1+3t_2}{4}\right]$ is within $260n^2a^3\varepsilon'$ and the variation on $\left[\frac{t_1+3t_2}{4}, t_2\right]$ is within $\frac{16m^2}{R}$. Notice that for $f \in F$,

$$\|\phi'(f)(t_1) - \phi'(f)(t_2)\| < \gamma,$$

$$\|\phi'(f)(t_i) - \phi(f)(t_i)\| < \gamma; \qquad i = 1, 2.$$

So for $t \in [t_1, t_2]$ and $f \in F$,

$$\begin{aligned}
&\|\phi'(f)(t) - \phi(f)(t)\| \\
&\leq \|\phi'(f)(t) - \phi'(f)(t_i)\| + \|\phi'(f)(t_i) - \phi(f)(t_i)\| \\
&\quad + \|\phi(f)(t_i) - \phi(f)(t)\| \\
&< \gamma + 260n^2a^3\varepsilon' + 16\frac{m^2}{R} + \gamma + \gamma \\
&< 300n^2a^3\varepsilon'
\end{aligned}$$

provided $\frac{1}{R} < \varepsilon'$ and $\gamma < \varepsilon'$. We may take ε' to be $\frac{\varepsilon}{300n^2a^3}$.

The relations of these numbers are as follows: $\frac{\delta \leq 1}{8m}$, $\delta \ll \gamma$, $2(2^a\delta + \frac{M}{N}) < \frac{1}{R^2}$, $\frac{1}{M} < \delta$, $\frac{M}{N} < \delta$, $\frac{1}{R} < \frac{3}{2}$, $\frac{1}{N} < \frac{\beta}{4}$, $\frac{16m^2}{R} < \gamma$, $36n^2m^2\frac{\gamma}{\beta} < 1$, $\beta \ll \varepsilon$, $\delta < \frac{1}{4m}$ and $1 < R < M < N$.

So for any $\varepsilon > 0$ and $F \subset A$, we form $\widetilde{F}$ (only depends on ε and the graphs X). Take $\beta \ll \varepsilon$ so that $\|f(s) - f(t)\| < \varepsilon'$ for $f \in F \cup \widetilde{F}$ whenever $d(s,t) < \beta$. Fix $\gamma < \frac{\beta}{36m^2n^2}$. Take R large enough so that $\frac{1}{R} < \frac{\beta}{2}$ and $\frac{1}{R} < \frac{\gamma}{16m^2}$. Once we have fixed R, take δ and $\frac{M}{N}$ so small that $2(2^a\delta + \frac{M}{N}) < \frac{1}{R^2}$. We may ask that $\delta \ll \gamma$ so that $\|f(s) - f(t)\| < \gamma$ if $d(s,t) < \delta$, for $f \in F \cup \widetilde{F}$. Increase M and N so that $\frac{M}{N} < \delta$, $\frac{1}{M} < \delta$ and $\frac{1}{N} < \frac{\beta}{4}$. Finally, we partition Y fine enough so that the variations of the images of the elements in $F \cup \widetilde{F}$ and $H \cup \widetilde{H}$, on each interval, after we have formed H and $\widetilde{H}$ for fixed $R < M < N$, under the map ϕ, are within γ and δ, respectively.

The map we needed has been defined on $[t_1, t_2]$.

Step 4.   Interpolate on the other parts.

Once we have defined ϕ' on $[t_1, t_2]$, the representation at t_2 is fixed. Choose any representation at t_3

$$\phi(f)(t_3) = Y \begin{bmatrix} f(s_1) & & \\ & \ddots & \\ & & f(s_{p'}) \end{bmatrix} Y^*.$$

As before, we may assume that these points have the distinctness property introduced at the beginning of the proof. Hence we may assume that $p' = p$. Replacing Y by another unitary $\widetilde{Y}$, we may assume that s_i and ξ_i are in one pair. Now we can carry out the construction we just did on $[t_2, t_3]$.

To complete the proof of the theorem, we must define ϕ' on the two intervals at the ends of the edge. We will construct ϕ' on the last small interval, say $[c, d]$. The construction on the first interval $[t_0, t_1]$ will be similar.

ϕ' was defined at c and d. In another words, the representations at both ends of $[c, d]$ are fixed. We will make ϕ' continuous at c while at d, we shall introduce some permutation. More precisely, write

$$\phi'(f)(d) = Y \begin{bmatrix} f(s_1) & & \\ & \ddots & \\ & & f(s_{p'}) \end{bmatrix} Y^*,$$

$$\phi'(f)(c) = W \begin{bmatrix} f(\xi_1) & & \\ & \ddots & \\ & & f(\xi_p) \end{bmatrix} W^*.$$

If we group some of the vertex points together, we may put $\{s_i\}_{i=1}^{p'}$ into the form $\{\dot{s}_i\}_{i=1}^{p}$ so that each of them is a single vertex point or a full vertex and so that they can be paired with $\{\xi_i\}_{i=1}^{p}$ to within $2(2^a\delta + \frac{4M}{N})$ one by one, as in Step 1. We may assume that $\dot{s}_i$ is paired with ξ_i.

Introduce a permutation X such that

$$Y \begin{bmatrix} f(s_1) & & \\ & \ddots & \\ & & f(s_{p'}) \end{bmatrix} Y^*$$

$$= YX \begin{bmatrix} f(\dot{s}_1) & & \\ & \ddots & \\ & & f(\dot{s}_p) \end{bmatrix} X^*Y^*.$$

On $\left[\frac{c+d}{2}, d\right]$, move $\{\xi_i\}_{i=1}^{p}$ to $\{\dot{s}_i\}_{i=1}^{p}$, keeping YX as the unitary; we then have defined ϕ' on $\left[\frac{c+d}{2}, d\right]$. Notice that we did not change ϕ' at d. To complete the proof, we should define ϕ' on $\left[c, \frac{c+d}{2}\right]$. But this is the same as Step 3. This concludes the proof of Theorem 3.1.

Remark : Let y_0 be a vertex of Y, the representations of ϕ' at y_0 coming from two edges of Y may be different. However, they only differ by a unitary permutation. The group of points $\{\xi_i\}_{i=1}^{p}$ associated with ϕ' is uniquely determined. If we replace a point in the two representations by some other point of X, the two new representations still give the same map.

The following corollary will not follow from Theorem 3.1 directly. However, it follows from the proof of the theorem. By passing to quotients, one can extend the theorem to maps from finite direct sums of basic building blocks to finite direct sums of basic building blocks.

Corollary.   Let $A = A_1 \oplus \ldots \oplus A_n$ be a finite direct sum of basic building blocks and let B be another basic building block with generic fibre M_m. For any unital $*$-homomorphism ϕ from A to B, any finite subset $F \subset A$ and any $\varepsilon > 0$, there exists a unital $*$-homomorphism ϕ' from A to B such that:

(1) $\|\phi(f) - \phi'(f)\| < \varepsilon$ for $f \in F$.

(2)   On each edge of Y, identified with $I = [0, 1]$,

$$\phi'(f_1 \oplus \ldots \oplus f_n)(t) = W(t) \begin{bmatrix} f_1(\alpha(t)) & & \\ & \ddots & \\ & & f_n(\xi(t)) \end{bmatrix} W^*(t)$$

for all $f_1 \oplus \ldots \oplus f_n \in A$ and all $t \in I$, where $W(\cdot)$ is a unitary in $M_m(C(I))$ and $\alpha(t), \ldots, \xi(t)$ arc continuous maps from I to the spectrum of A.

Chapter 4.   Approximate Intertwinings

This section is taken from [11].

4.1.   Let $A_1 \to A_2 \to \cdots$ and $B_1 \to B_2 \to \cdots$ be sequences of C*-algebras. Let us say that a sequence of C*-algebra homomorphisms $A_1 \to B_1 \to A_2 \to B_2 \to \cdots$ is an approximate intertwining of these two sequences if the diagram

$$
\begin{array}{ccccc}
A_1 & \longrightarrow & A_2 & \longrightarrow & \cdots \\
\downarrow & \nearrow & \downarrow & \nearrow & \\
B_1 & \longrightarrow & B_2 & \longrightarrow & \cdots
\end{array}
$$

is approximately commutative, in the sense that for any fixed element of any A_i (or B_i), the difference of the images of this element along two different paths in the diagram, starting at A_i (or B_i) and ending at the same place, converges to zero as the number of steps for which the two paths coincide, starting at the beginning, becomes large.

Theorem.   *Suppose that there exists an approximate intertwining of the sequences of C*-algebras $A_1 \to A_2 \to \cdots$ and $B_1 \to B_2 \to \cdots$. Then the inductive limit C*-algebras $A = \lim\limits_{\to} A_i$ and $B = \lim\limits_{\to} B_i$ are isomorphic.*

Proof: The map $A_i \to B$ obtained by going across n steps, down, and across converges as n tends to infinity. Hence one obtains a map $A \to B$, and in the same way a map $B \to A$. Since the map obtained by going across n steps, down, across any number of steps, and up and across converges to the identity on A as n tends to infinity, the composed map $A \to B \to A$ is the identity.

4.2 Theorem.   *Let $A_1 \to A_2 \to \cdots$ and $B_1 \to B_2 \to \cdots$ be sequences of separable C*-algebras. Let $a_1 = (a_{1,n})$, $a_2 = (a_{2,n}), \ldots$ and $b_1 = (b_{1,n})$, $b_2 = (b_{2,n}), \ldots$ be dense sequences in $A_1, A_2, \ldots$ and $B_1, B_2, \ldots$, respectively. Let $A_1 \to B_1 \to A_2 \to B_2 \to \cdots$ be a sequence of C*-algebra homomorphisms. For each $i = 1, 2, \ldots$ denote by S_i the finite subset of A_i consisting* *of the images in* A_i *of the first* i *terms of the sequences* $a_1, a_2, \ldots, a_{i-1}$ *and* $b_1, b_2, \ldots, b_{i-1}$, *along all possible paths in the diagram*

$$
\begin{array}{ccccc}
A_1 & \longrightarrow & A_2 & \longrightarrow & \cdots \\
\downarrow & \nearrow & \downarrow & \nearrow & \\
B_1 & \longrightarrow & B_2 & \longrightarrow & \cdots
\end{array}
$$

Similarly, denote by T_i *the finite subset of* B_i *consisting of the images in* B_i *of the first* i *terms of the sequences* $a_1, a_2, \ldots, a_i$ *and* $b_1, b_2, \ldots, b_{i-1}$, *along all possible paths in the diagram. Suppose that for each* i *and for each* $a \in S_i$, *the difference between the two images of* a *in* A_{i+1} *(along the two paths going straight across and down up) is of norm at most* 2^{-i}. *Suppose, similarly, that for each* i *and for each* $b \in T_i$, *the difference between the two images of* b *in* B_{i+1} *(along the two paths going straight across and up down) is of norm at most* 2^{-i}. *Then the sequence* $A_1 \to B_1 \to A_2 \to B_2 \to \cdots$ *is an approximate intertwining.*

Proof: For any two paths in the diagram, starting at A_i (or B_i) and ending at the same place, and coinciding until the jth stage (i.e., until they both pass through A_j, or both through B_j), and for any $n < j$, the images of $a_{i,n}$ (or $b_{i,n}$) along the two paths (which are equal to the images along the two paths continuing from B_j of some single element of T_j) differ by at most the sum of all discrepancies over triangles after the jth stage, i.e., by at most

$$
2(2^{-j} + 2^{-j-1} + \cdots) = 2^{-j+2}.
$$

This shows that the diagram is approximately commutative on any $a_{i,n} \in A_i$ (or $b_{i,n} \in B_i$). Since $\{a_{i,n} \mid n \in N\}$ is dense in A_i (and $\{b_{i,n} \mid n \in N\}$ is dense in B_i) the diagram is approximately commutative, as desired.

Chapter 5.   Asymptotic characterization

5.1.   Let X be a finite connected graph with multiple vertices. We can get a simple graph $\widetilde{X}$ out of X by identifying some of its points. More precisely, pick finitely many points in the interior of each edge of X, and identify all those points as well as those vertices with multiplicity one. The new graph we get, say $\widetilde{X}$, can be written as $\widetilde{X} = \widetilde{X}_1 \vee \widetilde{X}_2 \vee \cdots \vee \widetilde{X}_k$, i.e., k simple graphs $\{\widetilde{X}_i\}$ joined at a single point, say x_0. Each $\widetilde{X}_i$ is a circle or several edges connecting x_0 to another vertex x_i with multiple vertex points. The vertices of $\widetilde{X}$ can be divided into two groups. The first group is the point x_0 with multiplicity one. The second group consists of all those vertices of X at which there are multiple vertex points. We shall call any graph of this form a special graph. In 5.5 we will show that for a real rank zero C*-algebra that can be expressed as an inductive limit of a sequence of finite direct sums of basic building blocks, one can replace the sequence by another sequence so that each basic building block has a special graph as its spectrum. This is an important step in this program.

5.2.   Let A be a basic building block with generic fibre M_n and with non-Hausdorff spectrum X. Suppose X has k edges $L_1, L_2, \ldots, L_k$. We may identify each of them with the unit interval I. Here if L_i is a circle, we cut it at the vertex and identify it with I. Then we can embed A into $\bigoplus_{i=1}^{k} M_n\big(C(L_i)\big)$ in a natural way. We will denote this embedding by $\imath$. Now suppose that $V_i \in M_n\big(C(L_i)\big)$ is a unitary. We have an embedding of A into $\bigoplus_{i=1}^{k} M_n\big(C(L_i)\big)$ associated with the unitary $V = \mathrm{diag}(V_1, V_2, \ldots, V_k)$, i.e., the natural embedding followed by $Ad\,V$.

Definition.   Let A be a basic building block as above, let $F \subset A$ be a finite subset, and let $\varepsilon > 0$. We say F is approximately constant to within ε if there exists $V_i \in M_n\big(C(L_i)\big)$ such that for any $f \in F$, $(AdV) \circ \imath(f)$ is an almost constant matrix function with variation less than ε. Or equivalently, for each j

$$\big\|\big(Ad\,V_j(s)\big)f(s) - \big(Ad\,V_j(t)\big)f(t)\big\| < \varepsilon$$

29

for $s, t \in L_j$. We will say that F is approximately constant to within ε in the strong sense if for all $f \in F$ and for any k and j,

$$\left\| \left(Ad\, V_j(s) \right) f(s) - \left(Ad\, V_k(t) \right) f(t) \right\| < \varepsilon$$

for $s \in L_j$ and $t \in L_k$.

5.3 Theorem. Let A and B be two basic building blocks with generic fibres M_n and M_m and with spectra X and Y respectively. Let $F \subseteq A$ be a finite subset and let ϕ be a unital $*$-homomorphism from A to B. For any $\varepsilon > \delta > 0$, suppose that:

(1) For $f \in F$, $\|f(x) - f(x')\| < \varepsilon$ whenever $d(x, x') < 2^{a+5}\delta$, where a is the total number of vertex points of X.

(2) The eigenvalues of $\phi(h)(t)$ and the eigenvalues of $\phi(h)(s)$ are within δ one by one, in increasing order, for all the test functions h associated with a fixed positive number $M \geq 1$, in the sense of 2.6, and for any s and t in Y.

Then it follows that

(i) There is a sub-C^*-algebra $B^0 \subset B$, isomorphic to a basic building block with a special graph as its spectrum, and there exists a unital $*$-homomorphism ϕ_1 from A to B^0 such that

$$\|\phi(f) - \phi_1(f)\| < \varepsilon$$

for $f \in F$.

(ii) On each edge L of the spectrum of B^0, without the ends being identified, there exist a unitary $W(\cdot) \in M_m(C(L))$ and $\{\xi_i(\cdot)\}_{i=1}^p \subset C(L, X)$ such that

$$\phi_1(f)(t) \;=\; W(t) \begin{bmatrix} f(\xi_1(t)) & & \\ & \ddots & \\ & & f(\xi_p(t)) \end{bmatrix} W^*(t)$$

for $f \in A$ and $t \in L$. Furthermore, the variation of each of $\{\xi_i(t)\}_{i=1}^p$ is less than $2^{a+5}\delta$. As a consequence, $\phi_1(F)$ is approximately constant to within ε in B^0.

Proof: First we perturb ϕ so that it still satisfies (2) for 2δ and yet it has the form from the perturbation theorem.

For an edge L of Y, there are p maps $\{s_j(\cdot)\}_{j=1}^p$ in $C(L, X)$ associated with ϕ. P does not depends on the edges since the graph is connected.

Next we are going to pick some points on Y and then identify them so that we can get a special graph as in 5.1. Fix t_0 in Y such that

$$\phi(f)(t_0) \;=\; W(t_0) \begin{bmatrix} f(\xi_1) & & \\ & \ddots & \\ & & f(\xi_p) \end{bmatrix} W^*(t_0) \,.$$

We may assume that the points $\{\xi_i\}_{i=1}^p$ have the distinctness property. This can be achieved by taking t_0 to be a partition point in the proof of the perturbation theorem.

For any $t \in Y$ at which the fibre of B is M_m, we have

$$\phi(f)(t) \;=\; W(t) \begin{bmatrix} f(s_1(t)) & & \\ & \ddots & \\ & & f(s_p(t)) \end{bmatrix} W^*(t) \,.$$

By (2) and lemma 2.5, $\{s_i(t)\}_{i=1}^p$ and $\{\xi_i\}_{i=1}^p$ can be paired to within $2^{a+1}2\delta$ one by one. On each edge of Y, we are going to change these $\{s_i(t)\}_{i=1}^p$ a little such that we get p new maps and such that at some points of Y the points coming from X are exactly $\{\xi_i\}_{i=1}^p$. This will enable us to identify those points of Y to get a special graph.

Let us first look at a vertex $y_0 \in Y$ at which B has fibre M_m. Regard y_0 as a point on some edge we have

$$\phi(f)(y_0) \;=\; W(y_0) \begin{bmatrix} f(s_1(y_0)) & & \\ & \ddots & \\ & & f(s_p(y_0)) \end{bmatrix} W^*(y_0) \,.$$

Denote these points as $\{s_i\}_{i=1}^p$. We can pair them to within $2^{a+2}\delta$ with $\{\xi_i\}_{i=1}^p$, say s_i with ξ_{n_i}. Define a new map ϕ' at y_0 as follows:

$$\phi'(f)(y_0) \;=\; W(y_0) \begin{bmatrix} f(\xi_{n_1}) & & \\ & \ddots & \\ & & f(\xi_{n_p}) \end{bmatrix} W^*(y_0)$$

where $f \in A$.

This definition of ϕ' does not depend on the edge we choose. If we regard y_0 as an end point of another edge, the definition will give the same representation. This can be seen from the remark of 3.2.

In this way, we have defined ϕ' on those vertices of Y at which B have full fibres M_m. Next, we are going to extend it to whole of Y. It is enough to define ϕ' on an edge of Y identified with $[0,1]$. Without loss of generality, we may assume that B has full fibre M_m at 1 and smaller dimension fibre at 0.

Assume ϕ has the following representation on this edge:

$$\phi(f)(t) = W(t) \begin{bmatrix} f(s_1(t)) & & \\ & \ddots & \\ & & f(s_p(t)) \end{bmatrix} W^*(t) \,.$$

Fix a small number σ so that on $[0, 2\sigma]$ each of $\{s_i(t)\}_{i=1}^p$ varies within δ.

Cover $[\sigma, 1]$ with finitely many intervals such that in each interval there exists a common pairing of $\{s_i(t)\}_{i=1}^p$ with $\{\xi_i\}_{i=1}^p$. This can be done since $[\sigma, 1]$ is compact and $\{s_i(t)\}_{i=1}^p$ are continuous maps. Shrink these intervals so that each small interval intersects two others except at σ and 1. At each end, the small interval intersects only one other interval. We may shrink them again so that the middle point of each small interval is not in other intervals.

Denote the first interval by $[\sigma, \sigma + \sigma_1]$ for some $\sigma_1 < \sigma$. We can pair $\{\xi_i\}_{i=1}^p$ with $\{s_i(\sigma)\}_{i=1}^p$ one by one to within $2^{a+2}\delta$. Define $\tilde{\xi}_i(\sigma)$ to be one of the points in $\{\xi_i\}_{i=1}^p$, according to the pairing. Since

$$d\big(s_i(\sigma - \sigma_1),\ s_i(\sigma)\big) < \delta,$$

$$d\big(\tilde{\xi}_i(\sigma),\ s_i(\sigma)\big) < 2^{a+2}\delta,$$

we have

$$d\big(s_i(\sigma - \sigma_1),\ \tilde{\xi}_i(\sigma)\big) < 2^{a+3}\delta.$$

We first define $\{\tilde{\xi}_i(t)\}_{i=1}^p$ on $[0, \sigma]$ by

$$\tilde{\xi}_i(t) \;=\; \begin{cases} s_i(t) & t \in [0, \sigma - \sigma_1] \\ \tilde{\xi}_i(\sigma) & t = \sigma \\ \text{linear} & t \in [\sigma - \sigma_1, \sigma]. \end{cases}$$

Here, linear means moving from $\tilde{\xi}_i(\sigma - \sigma_1)$ to $\tilde{\xi}_i(\sigma)$ in a linear way. Now we have p maps $\{\tilde{\xi}_i(t)\}_{i=1}^p$ defined on $[0, \sigma]$. They agree at $t = 0$ with $\{s_i(t)\}_{i=1}^p$.

Denote the second interval by $[t_2', t_3]$ and denote the first interval $[\sigma, \sigma + \sigma_1]$ by $[t_1', t_2]$. We have the following picture

$$
\begin{array}{ccccccc}
-[- & -[- & -]- & -\cdot - & -]- & -\cdot - \\
\sigma & t_2' & t_2 & t_3' & t_3 & t_4
\end{array}
$$

where $[t_3', t_4]$ is the third interval. Define $\xi_i\left(\frac{t_2' + t_3}{2}\right)$ to be a point in $\{\xi_j\}_{j=1}^p$ according to the pairing associated with $[t_2', t_3]$, for $i = 1, 2, \ldots, p$. Extend them in a constant way until they meet the other two points t_2 and t_3'. We also extend these $\{\tilde{\xi}_i(\sigma)\}_{i=1}^p$ in a constant way from σ to t_2'. The next step is to define these maps on $[t_2', t_2]$. To do that, we will first show that $\tilde{\xi}_i(t_2')$ and $\tilde{\xi}_i(t_2)$ are close. To be specific, suppose that the two pairings associated with $[t_1, t_2]$ and $[t_2', t_3]$ are α and β, respectively. By our definition, $\tilde{\xi}_i(t_2) = \xi_{\alpha(i)}$ and $\tilde{\xi}_i(t_2') = \xi_{\beta(i)}$. By our construction,

$$d\big(\tilde{\xi}_i(t_2'),\ \tilde{\xi}_i(t_2)\big) < 2^{a+3}\delta.$$

For each i, interpolate $\tilde{\xi}_i(t)$ on $[t_2', t_2]$ by connecting them in a linear way. The variation of each of $\{\tilde{\xi}_i(t)\}_{i=1}^p$ is within $2^{a+3}\delta$.

By this way, we have defined $\{\tilde{\xi}_i(t)\}_{i=1}^p$ on $[0, t_3']$. It is clear that we can define them on $[0, 1]$ such that at the middle points and 1 they are $\{\xi_i\}_{i=1}^p$. We remark that the pairing we used for the last interval is the one we used to define ϕ' at 1. It is routine to check that

$$d\big(s_i(t),\ \tilde{\xi}_i(t)\big) < 2^{a+4}\delta.$$

For any $f \in A$, define

$$\phi'(f)(t) = W(t) \begin{bmatrix} f(\tilde{\xi}_i(t)) & & \\ & \ddots & \\ & & f(\tilde{\xi}_p(t)) \end{bmatrix} W^*(t)$$

for all t on this edge. This agrees with ϕ' at 0 and 1 defined before. Clearly, we can define ϕ' on whole Y the same way. For any t on Y and $f \in F$, we have

$$\|\phi(f)(t) - \phi'(f)(t)\|$$
$$\leq \max\{\|f(\tilde{\xi}_i(t)) - f(s_i(t))\|\} < \varepsilon.$$

We are now in a position to define a unital sub-C*-algebra B° of B. List those middle points as well as those vertices of Y at which B has full fibres M_m as $t_1, t_2, \ldots, t_b$. There exists a unitary $V_i \in U(m)$ such that

$$\phi'(f)(t_i) = V_i \begin{bmatrix} f(\xi_1) & & \\ & \ddots & \\ & & f(\xi_p) \end{bmatrix} V_i^*$$

for $i = 1, 2, \ldots, b$. Let

$$B^0 = \left\{ g \in B \mid V_i^* g(t_i) V_i = V_j^* g(t_j) V_j, \ i,j = 1, 2, \ldots, b \right\}.$$

It is easy to see that $B^0 \subset B$ is a unital sub-C*-algebra and $\phi'(A) \subseteq B^0$.

Finally, we show that B^0 is a basic building block with special spectrum and $\phi'(F)$ is approximately constant to within ε in B^0. Let

$$D = \left\{ g \in B \mid g(t_i) = g(t_j), \ i,j = 1, 2, \ldots, b \right\}.$$

At each vertex of Y where B has smaller size fibre we associate it with the unitary $I \in M_m$. Connect all these I and $\{V_i^*\}_{i=1}^b$ by a unitary path, we have a unitary $V \in B$. Clearly $Ad\,V$ gives an isomorphism of B^0 onto $(Ad\,V)(B^0)$. and

$$(Ad\,V^*)(B^0) = D.$$

We may regard D as a basic building block and its spectrum Y is X with $t_1, t_2, \ldots, t_b$ being identified.

To see that $\phi'(F)$ is approximately constant to within ε, list all the middle points as well as all the other vertices of Y, say $t_1, t_2, \ldots, t_{b'}$. Cut at these points to get finitely many intervals $\{I_i\}$. It is easy to see that the variations of $\{\tilde{\xi}_i(t)\}_{i=1}^p$ on each I_i are within $2^{a+5}\delta$. Since at one end of each interval the points are the same, i.e., $\{\xi_i\}_{i=1}^p$, the approximate constantness will be in the strong sense.

This completes the proof of the theorem.

Remark: The theorem can be extended to the general case: The first algebra is $A_1 \oplus \cdots \oplus A_k$ and the second algebra is $B_1 \oplus \cdots \oplus B_m$, where k and m are two positive integers and each of $\{A_i\}_{i=1}^k \cup \{B_i\}_{i=1}^m$ is a basic building block. It is easy to see from the proof of the theorem that the same proof works for the case that $k \geq 1$ and $m = 1$. In general, it can be reduced to the case $m = 1$ via the quotient map π_i from $B_1 \oplus \cdots \oplus B_m$ to B_i.

5.4. In this section we will prove the converse of Theorem 5.3. Clearly, the notion of eigenvalue variation for self-adjoint elements in a matrix algebra over a compact space defined in [4] can be extended to the basic building blocks in this paper in the same way.

In the case that the multiplicity for each vertex of the graphs is one, the following result can be found in [4] and [11]. The proof for our case only needs a minor modification. Notice that the condition (a) implies the condition (2) of Theorem 5.3 if we let A to be A_n and B to be A_m for some large m.

Theorem. Let $A = \lim_{\rightarrow} (A_n, \phi_{n,m})$ be a C^*-algebra which is the inductive limit of C^*-algebras A_n with unital $*$-homomorphisms $\{\phi_{n,m}\}$, where each A_n is a finite direct sum of basic building blocks. It follows that the following two properties are equivalent:

(a) For any self-adjoint element h in A and $\varepsilon > 0$, there is an $m > n$ such that the variation of the eigenvalues of $\phi_{n,m}(h)$ is less than ε.

(b) A is of real rank zero.

Proof: (b) $\Rightarrow$ (a) This was proved in [4] for the case that all the basic building blocks are homogeneous. The same proof works in this case.

(a) $\Rightarrow$ (b) For the case that all the basic building blocks are homogeneous, this was proved in [11]. The proof was reduced to the problem of approximating a self-adjoint element in a homogeneous basic building block by a self-adjoint element in the same algebra with distinct eigenvalues in each fibre. Clearly, this can be carried out in our case. One just perturbs the given self-adjoint element at vertices and then interpolates on each edge.

This completes the proof of the theorem.

The following proposition is the converse of Theorem 5.3. The special case that the graphs are circles and the basic building blocks are homogeneous was obtained in [11].

Proposition.   Let $A = \varinjlim (A_n, \phi_{n,m})$ be a C^*-algebra which is the inductive limit of C^*-algebras A_n with unital $*$-homomorphisms $\{\phi_{n,m}\}$, where each A_n is a finite direct sum of basic building blocks. The following two are equivalent:

(a) If for any n, for each finite subset F of A_n, and for $\varepsilon > 0$, there exists an $m > n$, a sub-C^*-algebra $A_m^0 \subset A_m$ isomorphic to a finite direct sum of basic building blocks, and a finite subset $F^0 \subset A_m^0$ such that the image of each element of F in A_m is within ε of an element in F^0 and such that the set of components of F^0 in each basic building block of A_m^0 is approximately constant to within ε in the strong sense.

(b) A is of real rank zero.

Proof: (b) implying (a) is Theorem 5.3. To prove that (a) implies (b), we will prove that the condition (a) in the proposition implies the condition (a) in Theorem 5.4. Notice that the converse is clearly true.

For $h = h^* \in A_n$, let $F = \{h\}$. There exists an $m > n$, a sub-C^*-algebra $A_m^0 \subset A_m$ isomorphic to a finite direct sum of basic building blocks, and an element $\tilde{h} \in F^0 = \{\tilde{h}\} \subset A_m^0$ such that $\|\phi_{n,m}(h) - \tilde{h}\| < \frac{\varepsilon}{4}$. Furthermore, $F^0 o \subset A_m^0$ is approximately constant to within $\frac{\varepsilon}{4}$ in the strong sense. Let ϕ be a $*$-isomorphism from A_m^0 onto $\phi(A_m^0)$ where $\phi(A_m^0)$ is a standard basic building block. By our assumption, $EV_{\phi(A_m^0)}(\phi(\tilde{h}))$ is less than $\frac{\varepsilon}{4}$. We would like to show that $EV_{A_m}(\tilde{h})$ is also less than $\frac{\varepsilon}{4}$. For convenience, we may assume that A_m is a single basic building block. ϕ^{-1} is a $*$-isomorphism from $\phi(A_m^0)$ into A_m and $\tilde{h} = \phi^{-1}(\phi(\tilde{h}))$. Approximate $\phi(\tilde{h})$ by another self-adjoint element k in $\phi(A_m^0)$ with finite spectrum, of multiplicity one, to within $\frac{\varepsilon}{4}$. Then $\phi^{-1}(k)$ has finite spectrum. Furthermore, $\|\tilde{h} - \phi^{-1}(k)\|$ is less than $\frac{\varepsilon}{4}$. This says that $\|\phi_{n,m}(h) - \phi^{-1}(k)\|$ is less than ε and hence $EV_{n,m}(\phi_{n,m}(h))$ is less than ε. This completes the proof of the Proposition.

5.5.   In this section we will show that arbitrary graphs can be replaced by the graphs with special forms in the sense of 5.1.

Theorem.   Let $B = \varinjlim (B_n, \phi_{n,m})$ be real rank zero inductive limit C^*-algebra, where each B_n is a finite direct sum of basic building blocks and each $\phi_{n,m}$ is a unital $*$-homomorphism. It follows that there exists a sequence of subindices $\{n_k\}_{k=1}^{\infty}$ and $A_{n_k} \subset B_{n_k}$, a C^*-algebra isomorphic to a finite direct sum of basic building blocks with special spectrum in the sense of 5.1, such that the C^*-algebra inductive limit $\varinjlim A_{n_k}$ is isomorphic to B.

Proof: The proof will be by constructing an approximately intertwining sequence in the sense of section 4.

Let $A_1 = B_1$ and $\{a_{1,n}\}_{n=1}^{\infty}$ be a dense sequence in A_1. Denote by $\{b_{1,n}\}_{n=1}^{\infty} \subset B_1$ the same sequence. Let $S_1 = \emptyset$ and $T_1 = \{b_{1,1}\}$. For $\varepsilon = \frac{1}{2}$, by Remark 5.3 there exist $n_2 > n_1 = 1$

and $A_{n_2} \subset B_{n_2}$, a finite direct sum of basic building blocks with special spectra, such that the following diagram commutes approximately on T_1 to within $\frac{1}{2}$:

$$
\begin{array}{ccc}
B_1 & \longrightarrow & B_{n_2} \\
 & \searrow & \uparrow \\
 & & A_{n_2}
\end{array}
$$

Fix a dense sequence $\{a_{2,m}\}_{m=1}^{\infty} \subset A_{n_2}$ and a dense sequence $\{b_{2,m}\}_{m=1}^{\infty} \subset B_{n_2}$. Let $S_2 \subset A_{n_2}$ consist of the images of the first two terms of $\{a_{1,n}\}_{n=1}^{\infty}$ and $\{b_{1,n}\}_{n=1}^{\infty}$ along all possible paths in the diagram

$$
\begin{array}{ccc}
B_{n_1} & \longrightarrow & B_{n_2} \\
\| & \searrow & \uparrow \\
A_{n_1} & \longrightarrow & A_{n_2},
\end{array}
$$

and let $T_2 \subset B_{n_2}$ denote the subset consisting of all the images of the first two terms of the sequences $\{a_{1,n}\}_{n=1}^{\infty}$, $\{a_{2,n}\}_{n=1}^{\infty}$ and $\{b_{1,n}\}_{n=1}^{\infty}$ along all possible paths in the above diagram. For $\varepsilon = \frac{1}{2^2}$, there exist $n_3 > n_2$ and $A_{n_3} \subset B_{n_3}$ as in Corollary 5.3 such that the following diagram approximately commutes on T_2 to within $\frac{1}{2^2}$ and the diagram commutes on A_{n_2}:

$$
\begin{array}{ccc}
B_{n_2} & \longrightarrow & B_{n_3} \\
\uparrow & \searrow & \uparrow \\
A_{n_2} & \longrightarrow & A_{n_3}
\end{array}
$$

Continuing this way, one has that the following approximately intertwining diagram

$$
\begin{array}{ccccccccc}
B_{n_1} & \longrightarrow & B_{n_2} & \longrightarrow & B_{n_3} & \longrightarrow & \cdots & \longrightarrow & B \\
\uparrow & \searrow & \uparrow & \searrow & \uparrow & & & & \\
A_{n_1} & \longrightarrow & A_{n_2} & \longrightarrow & A_{n_3} & \longrightarrow & \cdots & \longrightarrow & \varinjlim A_{n_k}
\end{array}
$$

satisfies the conditions of Theorem 4.2. As a consequence B is isomorphic to $\varinjlim A_{n_k}$.

Each A_{n_k} is isomorphic to a finite direct sum of standard basic building blocks via a $*$-isomorphism ϕ_n. This gives us the following commuting diagram:

$$
\begin{array}{ccccccccc}
A_{n_1} & \longrightarrow & A_{n_2} & \longrightarrow & A_{n_3} & \longrightarrow & \cdots & \longrightarrow & \varinjlim A_{n_k} \\
\updownarrow & & \updownarrow & & \updownarrow & & & & \\
\phi_1(A_{n_1}) & \longrightarrow & \phi_2(A_{n_2}) & \longrightarrow & \phi_3(A_{n_3}) & \longrightarrow & \cdots & \longrightarrow & \varinjlim \phi_k(A_{n_k})
\end{array}
$$

where the map from $\phi_k(A_{n_k})$ to $\phi_{k+1}(A_{n_{k+1}})$ is given by $\phi_{k+1} \circ \phi_{n_k,n_{k+1}} \circ \phi_k^{-1}$. This will ensure that $\varinjlim A_{n_k}$ and $\varinjlim \phi_k(A_{n_k})$ are isomorphic. Notice that the basic building blocks in each $\phi_k(A_{n_k})$ are standard ones with special spectra in the sense of 5.1.

Chapter 6.   Existence

In this section, we will lift maps from the K-groups to the corresponding basic building blocks.

6.1.   Let A be a basic building block with generic fibre M_n and with special spectrum X. As a graph with special form, X is formed by $q + r$ parts joined at a vertex x_0 with multiplicity one, i.e., $X = X_1 \vee X_2 \vee \ldots \vee X_q \vee \ldots \vee X_{q+r}$, where each of the first q parts has another vertex with multiplicity at least two and the last r parts are circles. For $1 \leq i \leq q$, denote the other vertex of X by $x_i = \{x_{ij}\}_{j=1}^{n_i}$ and denote the fibre at x_{ij} by $M_{m_{ij}}$. Notice that there are some edges, say k_i of them, that connect x_0 and x_i in X_i.

To this basic building block, one can associate a four-term exact sequence. More precisely, let I be the ideal of all the elements of A vanishing at all the vertices of X. By Bott periodicity [2], one has the following six-term exact sequence

$$
\begin{array}{ccccc}
K_0(I) & \longrightarrow & K_0(A) & \longrightarrow & K_0(A/I) \\
\uparrow & & & & \downarrow \\
K_1(A/I) & \longleftarrow & K_1(A) & \longleftarrow & K_1(I)
\end{array}
$$

To reduce this to a four-term exact sequence, notice first that $K_0(I)$ and $K_1(A/I)$ are trivial. This is because the quotient algebra A/I can be identified with a finite dimensional algebra

$$
M_n \oplus \Big(\bigoplus_{j=1}^{n_1} M_{m_{1j}} \Big) \oplus \cdots \oplus \Big(\bigoplus_{j=1}^{n_q} M_{m_{n_qj}} \Big) .
$$

Hence

$$
K_0(A/I) \ = \ \mathbb{Z} \oplus \mathbb{Z}^{n_1} \oplus \cdots \oplus \mathbb{Z}^{n_q}
$$

with the usual order. Since

$$
I \cong M_n \Big(\bigoplus^{k_1} C_0(T) \Big) \oplus \cdots \oplus \Big(\bigoplus^{k_q} C_0(T) \Big) \oplus \Big(\bigoplus^{r} C_0(T) \Big)
$$

one also has $K_1(I) \cong \mathbb{Z}^{k_1 + \cdots + k_q + r}$.

The six-term exact sequence is hence reduced to

$$0 \longrightarrow K_0(A) \xrightarrow{\pi_*} K_0(A/I) \xrightarrow{\partial_A} K_1(I) \xrightarrow{\iota_*} K_1(A) \longrightarrow 0$$

where π is the quotient map, ∂_A is the so-called exponential map and ι is the inclusion map. Identify each edge of the first q parts $\{X_i\}_{i=1}^q$ with $[0,1]$, from x_0 to x_i. We will call this direction positive. Namely, we will assign the class of the following unitary one in $K_1\big(C_0(L)\big)$ for an edge L:

$$u = \begin{cases} \begin{bmatrix} e^{2\pi i t} & & & & \\ & 1 & & & \\ & & 1 & & \\ & & & \ddots & \\ & & & & 1 \end{bmatrix} & t \in L \\[2em] I & t \in X\backslash L. \end{cases}$$

Under this assumption, the exponential map ∂_A, as defined in [2] in general, can be identified by the following matrix:

$$\partial_A = \begin{array}{c} k_1 \left\{ \begin{array}{c} \\ \\ \\ \end{array} \right. \\ \vdots \\ k_q \left\{ \begin{array}{c} \\ \\ \\ \end{array} \right. \\ r \left\{ \begin{array}{c} \\ \\ \end{array} \right. \end{array} \begin{bmatrix} \overbrace{\quad}^{n_1} & \overbrace{\quad}^{n_2} & \overbrace{\quad}^{n_q} \\ 1 & -1\cdots-1 & & \\ \vdots & \cdots & & \\ 1 & -1\cdots-1 & & \\ 1 & & -1\cdots-1 & \\ \vdots & & \cdots & \\ 1 & & -1\cdots-1 & \\ & & & \ddots \\ 1 & & & -1\cdots-1 \\ \vdots & & & \cdots \\ 1 & & & -1\cdots-1 \\ 0 & & \cdots & 0\cdots0 \\ \vdots & & & \cdots \\ 0 & & \cdots & 0\cdots0 \end{bmatrix}.$$

Notice that it is not necessary to fix a direction for the circle parts. But we fix a direction for convenience.

We are now ready to compute $K_0(A)$ and $K_1(A)$. In general, we have the following:

$$K_0(A) \cong \operatorname{Ker} \partial_A$$

and

$$K_1(A) \cong K_1(I)\Big/\operatorname{Im} \partial_A$$

Let $z = (z_0, z_{11}, \ldots, z_{1n_1}, z_{21}, \ldots, z_{2n_2}, \ldots, z_{q_1}, \ldots, z_{qn_q})^{tr} \in K_0(A/I)$; then the equation $\partial_A z = 0$ is equivalent to the following system:

$$z_0 = \sum_{j=1}^{n_1} z_{1j} = \sum_{j=1}^{n_2} z_{2j} = \cdots = \sum_{j=1}^{n_q} z_{qj}.$$

In fact, we can realize this group in terms of the projections in A. It is easy to see that the positive cone consists of all the elements with non-negative entries. For convenience, we will say that an element of $K_0(A/I)$ has $q+1$ blocks: $z_0, z_1 = (z_{11}, \ldots, z_{1n_1}), \ldots, z_q = (z_{q1} \cdots z_{qn_q})$.

6.2 Lemma. *Let $E \subset K_0^+(A)$ be the subset consisting of all the elements with one entry one and other entries zero in each block. Then E is a generating set for $K_0^+(A)$ and $K_0(A)$.*

Proof: Let $z' = (z_0', z_1', \ldots, z_q')^{tr} \in E \subset K_0^+(A)$; then $z_0' = 1$ and there is a one in each z_i' and all the other entries in z_i' are zero. Take $z = (z_0, z_{11}, \ldots, z_{1n_1}, \ldots, z_{q1} \cdots z_{qn_q})^{tr} \in K_0^+(A)$; then $z_0 > 0$ and $z_{ij} \geq 0$. Notice that $\sum_{j=1}^{n_i} z_{ij} = z_0$; we then can find $e \in E$ such that $z - e \in K_0^+(A)$. Continuing this way, we can express z as a sum of elements from E.

For $z = (z_0, z_1, \ldots, z_q)^{tr} \in K_0(A)$, take any $e \in E$; then $z - z_0 e \in K_0(A)$. The first entry of this element is zero. So we may assume that $z_0 = 0$. Now

$$\sum_{j=1}^{n_1} z_{1j} = \sum_{j=1}^{n_2} z_{2j} = \cdots = \sum_{j=1}^{n_q} z_{qj} = 0.$$

To make the second block zero, first notice that the sum of the positive entries is the same as the sum of the negative entries multiplied by negative one. For example, assume $z_{1j} > 0$ and $z_{1k} < 0$ and let $e_1, e_2 \in E$ be such that e_1 and e_2 are only different in the second block. Namely, e_1 has one in the jth place while e_2 has one in the kth plane. Clearly $z - (e_1 - e_2) \in K_0(A)$. Now the sum of the positive entries of $z - (e_1 - e_2)$ is less than that of z. Continuing this way, we can eventually make the second block zero. After we go through all blocks, we finally express z as a combination of the elements of E. (In fact, this fact is obvious in our case since $K_0^+(A) - K_0^+(A) = K_0(A)$.)

6.3 Lemma. *For $i = 1, \ldots, q$, let $\varepsilon^{(i)} = (\varepsilon_0, \ldots, \varepsilon_q) \in K_0^+(A/I)$ be as follows: all the blocks ε_j are zero except ε_i where $\varepsilon_i = (1, 0, 0, \ldots, 0)$. Then the group G generated by these q elements has the property that $G \oplus K_0(A) = K_0(A/I)$.*

Proof: For any $z = (z_0, z_1, \ldots, z_1)^{tr} \in K_0(A/I)$, let $a_i = \sum_{j=1}^{n_i} z_{ij} - z_0$. Then $z - a_1\varepsilon_1 - \cdots - a_q\varepsilon_q \in K_0(A)$. This shows that $G + K_0(A) = K_0(A/I)$.

Suppose $z = (z_0, z_1, \ldots, z_q)^{tr} \in K_0(A) \cap G$, say

$$z = a_1\varepsilon_1 + a_2\varepsilon_2 + \cdots + a_q\varepsilon_q.$$

Then

$$\begin{cases} z_0 = 0 \\ z_{11} = a_1, \quad z_{12} = \cdots = z_{1n_1} = 0 \\ \cdots \\ z_{q1} = a_q, \quad z_{q2} = \cdots = z_{qn_q} = 0. \end{cases}$$

But $z = (0, a_1, 0, \cdots, 0, a_2, 0, \cdots, 0, \cdots, a_q, 0, \cdots, 0)^{tr} \in K_0(A)$ if and only if $a_1 = a_2 = \cdots = a_q = 0$. This says that the intersection of G and $K_0(A)$ is trivial.

6.4. Let B be another basic building block with spectrum X' and generic fibre $M_{n'}$. Suppose that X' has special form with special vertex x_0' and q' other vertices $\{x_i'\}_{i=1}^{q'}$. We write $X' = X_1' \vee \ldots \vee X_{q'}' \vee \ldots \vee X_{q'+r'}'$, where each X_i' has k_i' edges and at x_i' there are n_i' points $x_{i1}', \ldots, x_{in_i'}'$, $i = 1, 2, \ldots, q'$. Assume that the fibre at x_{ij}' is $M_{m_{ij}'}$. Denote by J the ideal of B of those elements vanishing at all vertices.

Lemma. *Let A and B be as above and let ϕ_0 be a map from $K_0(A)$ to $K_0(B)$, preserving the order structure. Then there is an extension ϕ of ϕ_0 from $K_0(A/I)$ to $K_0(B/J)$, mapping the positive cone into the positive cone.*

Proof: Let G_A and G_B be the groups in Lemma 6.3 for A and B, respectively. It is easy to define a map ϕ from $G_A \oplus K_0(A)$ to $G_B \oplus K_0(B)$ such that the restriction on $K_0(A)$ is ϕ_0. We only have to send those generators of G_A to G_B. ϕ is a map from $Z^{1+\alpha}$ to $Z^{1+\beta}$ where $\alpha = n_1 + \ldots + n_q$ and $\beta = n_1' + \ldots + n_{q'}'$. In the standard basis ϕ has the following representation:

$$\phi = \begin{bmatrix} \phi_{00}, \phi_{01} \ldots \phi_{0n_1}, & \cdots & \phi_{0\alpha} \\ \phi_{10}, \phi_{11} \ldots \phi_{1n_1}, & \cdots & \phi_{1\beta} \\ & & \\ \phi_{\beta 0}, \phi_{\beta 1} \ldots \phi_{\beta n_1}, & \cdots & \phi_{\beta\alpha} \end{bmatrix}.$$

Since $\{\varepsilon^{(i)}\} \cup E$ may not generate the positive cone, we can not expect the entries of ϕ to be non-negative.

Let ψ be a $(1 + \beta) \times (1 + \alpha)$ matrix. If $\psi(K_0(A)) = 0$, $\phi + \psi$ will be another extension of ϕ_0. Write the first row of ψ as

$$\psi_0 = (\psi_{00}, \psi_{01}, \ldots \psi_{0n_1}, \ldots, \ldots \psi_{0\alpha}).$$

For $z = (0, 1, -1, 0 \ldots 0)^{tr} \in K_0(A)$, $\psi_0 \cdot z = 0$ gives us $\psi_{01} = \psi_{02}$. It is easy to see now that $\psi(K_0(A)) = 0$ implies that the entries in each block are the same. If we apply ψ to $z = (1, 1, 0, \cdots, 0, 1, 0, \cdots, 0, \ldots, 1, 0, \ldots)^{tr} \in K_0(A)$, i.e., the first entry of each block is one, then

$$\psi_{00} + \psi_{01} + \psi_{0n_1+1} + \cdots + \psi_{0\alpha} = 0 .$$

The similar result holds for each row of ψ.

The converse is also true. If each row of ψ has the property that all the entries in the same block are the same and the sum of these $q + 1$ numbers is zero, then $\psi(E) = 0$. Hence $\psi(K_0(A)) = 0$.

We are now in a position to change ϕ so that its representation matrix has non-negative entries. First, we claim that for any fixed row, say the first row, there is at least one block with all positive entries unless this row is already non-negative. To see that, fix a position in each block at which the entry has smallest value, and choose an $e \in E$ such that the entries at those positions are one. Since $\phi(e) = \phi_0(e) \in K_0^+(B)$, this says that the sum of those numbers is non-negative. If there is at least one block with negative entry, there must be one number positive. The block this number belongs to has all positive entries. Divide the $q + 1$ blocks of this row into two groups, the first consisting of all those blocks whose entries are positive, and the second containing all others. Let ψ be defined as follows. For $i \geq 1$, if the ith block has all positive entries with a_i the smallest one, then we put $-a_i$ across this block for ψ. Otherwise, we simply put zero. At the first position, we put the sum of all those a_i. For all other rows, enter the zero entry. It is clear that $\psi(K_0(A)) = 0$. So $\phi + \psi$ is an extension of ϕ. We will still denote this map by ϕ.

By our claim above, there exists at least one block in the first row such that all its entries are positive. Clearly, this must be the first block, ϕ_{00}. In other words, the first group consists of only this block. To turn the other blocks into the first group, let us first examine the second block $(\phi_{01}\phi_{02}, \ldots, \phi_{0n})$. We may assume that $a = \min(a_{01}, \ldots, a_{0n_1}) < 0$. Let

$$\psi = \begin{bmatrix} a, -a, & \ldots & -a, 0 \cdots 0 \\ 0, & 0, & \ldots & 0 \cdots 0 \\ & & \ldots & \\ 0 & & \ldots & 0 \end{bmatrix},$$

i.e., the first block is a and the second block is $-a > 0$ across. Now $\phi + \psi$ is an extension of ϕ_0 and the second block becomes non-negative. The smallest entry is zero. Notice that unless each block of the first row of $\phi + \psi$ is non-negative, in which case we are done, the first entry of this row is still positive. So we are ready to turn another block with negative entries into non-negative one.

Continuing in this way, we can turn ϕ into a non-negative matrix which is still an extension of ϕ_0. This completes the proof of the lemma.

6.5.   Suppose that we have two basic building blocks A and B as above, and suppose that two maps ϕ_0 and ϕ_1 from $K_0(A)$ to $K_0(B)$ and from $K_1(A)$ to $K_1(B)$ are given. By Lemma 6.4, we can extend ϕ_0 to ϕ and hence we have the following diagram:

$$
\begin{array}{ccccccccccc}
0 & \to & K_0(A) & \to & K_0(A/I) & \to & K_1(I) & \to & K_1(A) & \to & 0 \\
 & & \phi_0 \downarrow & & \phi \downarrow & & & & \phi_1 \downarrow & & \\
0 & \to & K_0(B) & \to & K_0(B/J) & \to & K_1(J) & \to & K_1(B) & \to & 0
\end{array}
$$

where the two rows are exact and the square at the left end commutes.

Since $K_1(I) \cong \mathbb{Z}^{(\sum_{i=1}^{q} k_i)+r}$ is projective, there is a group homomorphism η from $K_1(I)$ to $K_1(J)$ such that the following diagram commutes:

$$
\begin{array}{ccccccccccc}
0 & \to & K_0(A) & \to & K_0(A/I) & \to & K_1(I) & \to & K_1(A) & \to & 0 \\
 & & & & & & \eta \downarrow & & \phi_1 \downarrow & & \\
0 & \to & K_0(B) & \to & K_0(B/J) & \to & K_1(J) & \to & K_1(B) & \to & 0.
\end{array}
$$

Lemma.   In the standard bases of $K_1(I)$ and $K_1(J)$, η can be written as $\eta = (\eta_1, \ldots, \eta_q, \eta_{q+1} \ldots \eta_{q+r})$ where each η_i is a $\left(\sum_{i=1}^{q'} k_i' + r'\right) \times k_i$ matrix for $1 \le i \le q$ and where the last r matrices are all $\left(\sum_{i=1}^{q'} k_i' + r'\right) \times 1$. We can choose η in such a way that the sum of each row of η_i is zero for $i = 1, 2, \ldots, q$.

Proof: Denote the maps from $K_0(A)$ to $K_0(A/I)$, from $K_0(A/I)$ to $K_1(I)$ and from $K_1(I)$ to $K_1(A)$ by π_A, ∂_A and $\imath_A$ respectively. Similarly, we have the maps π_B, ∂_B and $\imath_B$.

Let $\bar{\eta}$ be a map from $K_1(I)$ to $K_1(J)$ such that $\imath_B \circ \bar{\eta} = 0$. Then $\eta + \bar{\eta}$ is also a lifting of ϕ_1. Since the second row is exact, $\imath_B \bar{\eta} = 0$ if and only if the image of $\bar{\eta}$ is in the image of ∂_B. For any $z \in K_0(B/J)$, z can be written as $z = (z_0, z_{11}, \ldots, z_{1n_1'} \ldots z_{q'n_{q'}'})^{tr}$. So

$$
\partial_B \begin{bmatrix} z_0 \\ z_{11} \\ \vdots \\ z_{q'n_{q'}'} \end{bmatrix} = \begin{bmatrix} z_0 - (z_{11} + \cdots + z_{1n_1'}) \\ \vdots \\ z_0 - (z_{11} + \cdots + z_{1n_1'}) \\ \vdots \\ z_0 - (z_{q'1} + \cdots + z_{q'n_q'}) \\ 0 \cdots \quad \cdots \quad 0 \\ 0 \cdots \\ 0 \cdots \quad \cdots \quad 0 \end{bmatrix} \begin{array}{l} \left.\right\} k_1' \\ \\ \\ \left.\right\} r' \end{array} \quad ,
$$

i.e., this image is of the form

$$
\big(\underbrace{a_1 \ldots a_1}_{k_1'}, \underbrace{a_2 \ldots a_2}_{k_2'} \cdots \underbrace{a_{q'} \ldots a_{q'}}_{k_{q'}'}, \underbrace{0 \ldots 0}_{r'}\big)^{tr}.
$$

Conversely, any element of this form is in the image of ∂_B. This can be seen by taking

$$\Big(0, \underbrace{-a_1, 0 \cdots 0}_{k_1'}, \underbrace{-a_2, 0 \cdots 0}_{k_2'} \cdots \underbrace{-a_{q'}, 0 \cdots 0}_{k_{q'}'} \underbrace{0 \cdots 0}_{r'}\Big)^{tr} \in K_0(B/J) .$$

Hence, $\pi_B \bar{\eta} = 0$ if and only if each column of $\bar{\eta}$ is of this form.

Consider $z = (1, 1, \ldots, 1, 0, 0 \cdots 0)^{tr} \in K_1(I)$; the first k_1 entries are one and all the other entries are zero. Clearly,

$$z = \partial_A (0, -1, 0, 0, , \ldots, 0)^{tr}$$

where $(0, -1, 0, \ldots)^{tr} \in K_0(A/I)$. Computing

$$\imath_B \eta(z) = \phi_1 \imath_A(z) = \phi_1 \imath_A \partial_A (0, -1, 0, \ldots)^{tr} = 0,$$

we conclude that $\eta(z) \in Im \partial_B$. Let

$$\bar{\eta}_1 = \eta(z) = (a_{11} \ldots a_{11}, a_{21} \ldots a_{21}, \ldots a_{q'1} \ldots a_{q'1}, 0 \cdots 0)^{tr} .$$

Those numbers are the sums of the first k_1 entries of each row of η. Similarly, we can get $\bar{\eta}_2, \ldots, \bar{\eta}_q$ corresponding to the second, third, ... and q^{th} sections of η.

Construct a matrix $\bar{\eta}$ as follows:

$$\bar{\eta} = \Big(\underbrace{-\bar{\eta}_1, 0, \cdots, 0}_{k_1}, \underbrace{-\bar{\eta}_2, 0, 0, \cdots, 0}_{k_2}, \ldots, \underbrace{-\bar{\eta}_q, 0, \cdots, 0}_{k_q}, \underbrace{0, \cdots, 0}_{r}\Big) .$$

This is a $\big(\sum\limits_{i=1}^{q'} k_i' + r'\big) \times \big(\sum\limits_{i=1}^{q} k_i + r\big)$ matrix and $\pi_B \bar{\eta} = 0$. Now $\eta + \bar{\eta}$ has the required property. This completes the proof of the lemma.

6.6. Theorem.   *Let A and B be two basic building blocks with special spectra and let ϕ_0 and ϕ_1 be two group homomorphisms from $K_0(A)$ to $K_0(B)$ and from $K_1(A)$ to $K_1(B)$, respectively. If $\phi_0([1_A]) = [1_B]$ and $\phi_0 \oplus \phi_1$ preserves the dimension range, then there exists a unital $*$-homomorphism ψ from A to B such that $K_*(\psi) = \phi_0 \oplus \phi_1$.*

Proof: Recall that for a unital $*$-homomorphism from A to B one can always associate with it a graded group homomorphism from $K_*(A)$ to $K_*(B)$ which preserves dimension range (see 1.1). Theorem 6.6 says that the converse is also true.

What we are going to do is to define a map ψ from A to B on each part of X' such that the definition agrees at x_0'. It is sufficient to work on the first part of X'. We would like to point out that although X_1' is not a circle the construction on a circle of X' will be the same. This will be clear from the proof. In this proof, we will use the previous notations used in 6.1 and 6.4.

Recall that the first part of X' has two vertices x'_0 and x'_1. At x'_1, there are n'_1 points $x'_{11}, x'_{12}, \ldots, x'_{1n'_1}$. In this part, there are k'_1 edges $L_1, L_2, \ldots, L_{k'_1}$. Identify each of these with $[0, 1]$ from x'_0 to x'_1.

By Lemma 6.4 and Lemma 6.5, there are two maps ϕ and η from $K_0(A/I)$ to $K_0(B/J)$ and from $K_1(I)$ to $K_1(J)$ respectively such that the following diagram commutes on the two squares at the both ends:

$$
\begin{array}{ccccccccc}
0 & \to & K_0(A) & \to & K_0(A/I) & \to & K_1(I) & \to & K_1(A) & \to & 0 \\
 & & \phi_0 \downarrow & & \phi \downarrow & & \eta \downarrow & & \phi_1 \downarrow & & \\
0 & \to & K_0(B) & \to & K_0(B/J) & \to & K_1(J) & \to & K_1(B) & \to & 0.
\end{array}
$$

The map ϕ can be lifted to a unital $*$-homomorphism F from A/I to B/J. The quotient map from A to A/I followed by F gives a map from A to B/J. We may still call this map F. The map from A/I to B/J can be chosen to have diagonal form.

We first define ψ at each vertex of X'. For a vertex x'_i of X' and any $f \in A$, define

$$
\psi(f)(x'_i) = F_i(f)
$$

where F_i is the map F from A to B/J followed by the quotient map from B/J to the summand x'_i represents. Notice that this representation has diagonal form. Next, we are going to interpolate the map on each edge of X'.

Write ϕ as the following matrix:

$$
\phi = \begin{bmatrix}
\phi_{00}, \phi_{01} \ldots \phi_{0n_1} & \cdots & \phi_{0\alpha} \\
\phi_{10}, \phi_{11} \ldots \phi_{1n_1} & \cdots & \phi_{1\alpha} \\
 & \cdots & \\
\phi_{\beta 0}, \phi_{\beta 1} \ldots \phi_{\beta n_1} & \cdots & \phi_{\beta\alpha}
\end{bmatrix}.
$$

Let $z \in K_0(A)$ be the element whose first two entries of its second block are $(1, -1)$ and all the other entries are zero, i.e., $z = (0, 1, -1, 0, \ldots 0)^{tr}$. The equation

$$
\partial_B \phi(z) = \partial_B \circ \imath_B \circ \phi_0(z) = 0
$$

implies that

$$
\phi_{01} - (\phi_{11} + \phi_{21} + \cdots + \phi_{n'_1 1}) = \phi_{02} - (\phi_{12} + \phi_{22} + \cdots + \phi_{n'_1 2}).
$$

It is easy to see that the same argument shows that

$$
\phi_{01} - (\phi_{11} + \phi_{21} + \cdots + \phi_{n'_1 1}) = \phi_{0i} - (\phi_{1i} + \phi_{2i} + \cdots + \phi_{n'_1 i}).
$$

for $1 \leq i \leq n_1$. We denote this common difference by a_1. Similarly, the same result holds in other blocks. So we have $a_1, a_2, \ldots, a_q$. We can define a_0 directly since there is only one column. For any $e \in E \subset K_0^+(A)$, the equation $\partial_B \phi(e) = 0$ gives us

$$\sum_{i=0}^{q} a_i = 0.$$

Let us examine the meaning of these $\{a_i\}_{i=0}^{q}$. We may regard ϕ as a correspondence from the vertex points of X' to the vertex points of X. For example, $a_1 > 0$ means that ϕ places a_1 more $x_1 = \{x_{1i}\}_{i=1}^{n_1}$ at x_0' than at x_1'. Notice that the difference is a_1 of the whole group $x_1 = \{x_{1i}\}_{i=1}^{n_1}$.

With this observation in mind, we are ready to extend ϕ. For the representations of ϕ at x_0' and x_1', there are two groups of the vertices of X corresponding to them respectively. Denote the vertices correspond to ϕ at x_0' by $\{\xi_i\}$ and denote the vertices correspond to ϕ at x_1' by $\{\lambda_i\}$. We now pair these vertices from the two groups. For the points in $\{\xi_i\} \cap \{\lambda_i\}$, one in $\{\xi_i\}$ pairs with the same one in $\{\lambda_i\}$. To pair the remaining ones, we notice that they can be grouped into vertices of X (as we observed above). Since $\sum_{i=0}^{q} a_i = 0$, the numbers of the whole vertices of X at x_0' and at x_1' should be the same. So we just pair a whole vertex of $\{\xi_i\} \backslash \{\xi_i\} \cap \{\lambda_i\}$ with a whole vertex of $\{\lambda_i\} \backslash \{\xi_i\} \cap \{\lambda_i\}$.

For a pair (ξ_i, λ_j), we define a map from X_1' to X. Recall that X_1' has k_1' edges $L_1, L_2, \ldots, L_{k_1'}$. Identify each of these with $[0,1]$ from x_0' to x_1'. There are two cases.

(a) $\xi_i = \lambda_j$. In this case define the map as

$$\lambda(\xi_i, \lambda_j, t) = \xi_i$$

for $t \in L_1 \vee L_2 \vee \cdots \vee L_{k_1'} = X_1'$. Here ξ_i could be a full vertex or a vertex point. Notice that we have actually defined k_1' maps, one on each edge of X_1'.

(b) $\xi_i \neq \lambda_j$. In this case, ξ_i and λ_j must be two different full vertices. We may denote ξ_i by x_i and λ_j by x_j. We consider the case that both of them are not x_0. The construction would be the same if one of them is x_0. Identify the two connected edges of X from x_j to x_i, via x_0, with $[0,2]$. Define

$$\lambda(\xi_i, \lambda_j, t) = 2 - 2t$$

for $t \in [0,1]$ identified with L_i for $i = 1, 2, \ldots, k_1'$.

Once we have these maps, we can define a map on the algebra level. Write two representations of ψ at x_0' and at x_i' as follows:

$$\psi(f)(x_0') \;=\; F_0(f) \;=\; \begin{bmatrix} f(\xi_1) & & & \\ & f(\xi_2) & & \\ & & \ddots & \end{bmatrix},$$

$$\psi(f)(x_1') \;=\; F_1(f) \;=\; \begin{bmatrix} f(\lambda_1) & & & \\ & f(\lambda_2) & & \\ & & \ddots & \end{bmatrix}.$$

Let U be a permutation matrix such that

$$F_1(f) \;=\; U \begin{bmatrix} f(\lambda_{j_1}) & & & \\ & f(\lambda_{j_2}) & & \\ & & \ddots & \end{bmatrix} U^*$$

where $\{\xi_p, \lambda_{j_p}\}$ is a pair for all p. Connect I to U on $[0,1]$ by a unitary path $W(t)$ in $M_{n'}(C[0,1])$. Define

$$R(f)(t) \;=\; W(t) \begin{bmatrix} f(\lambda(\xi_i, \lambda_{j_1}, t)) & & & \\ & f(\lambda(\xi_2, \lambda_{j_2}, t)) & & \\ & & \ddots & \end{bmatrix} W^*(t)$$

for $t \in L_1 \vee L_2 \vee \cdots \vee L_{k_1'}$. Clearly, we have

$$R(f)(0) \;=\; F_0(f)\,, \qquad R(f)(1) \;=\; F_1(f)$$

for all $f \in A$. In this way, we have defined a map from A to B restricted on X_1'.

In fact, this map is not what we want. It is easy to see that its K_1 contribution is zero. We are going to put in the information coming from η. The modification can be divided into two cases.

Case 1. At least one block of the first row of ϕ has all positive entries.

Recall that the first row is as follows:

$$(\phi_{00}, \phi_{01}, \ldots, \phi_{0n_1}, \ldots, \phi_{0\alpha})\,.$$

We may assume that $(\phi_{01}, \ldots, \phi_{0n_1})$ are all positive. So we have a $\lambda(\xi_i, \lambda_{j_i}, t)$ such that $\lambda(\xi_i, \lambda_{j_i}, 0) = x_1$ or we can find n_1 of $\lambda(\xi_i, \lambda_{j_i}, t)$ such that they form a full vertex x_1. In the later case, we will denote this whole group of maps by $\lambda(\xi_i, \lambda_{j_i}, t)$. Next, we insert the information from the first row of η into $\lambda(\xi_i, \lambda_{j_i}, t)$ for $t \in L_1$. Write the first row of η as:

$$(\eta_{11}, \ldots, \eta_{1k_1}, \eta_{21}, \ldots, \eta_{2k_2}, \ldots, \eta_{q_1}, \ldots, \eta_{qk_q}, \eta_1, \ldots, \eta_r)\,.$$

Recall that $\sum_{j=1}^{k_i} \eta_{ij} = 0$ for $1 \leq i \leq q$. Define a map $S(\xi_i, \lambda_{j_i}, t)$ from $L_1 = [0,1]$ to X as follows. On $[0, \frac{1}{4}]$, $S(\xi_i, \lambda_{j_i}, t)$ goes from x_1 to x_0. On $[\frac{3}{4}, 1]$, $S(\xi_i, \lambda_{j_i}, t)$ goes from x_0 to λ_{j_i}. $S(\xi_i, \lambda_{j_i}, t)$ and $\lambda(\xi_i, \lambda_{j_i}, t)$ agree at $t = 0$ and at $t = 1$.

It remains to fill in $[\frac{1}{4}, \frac{3}{4}]$. Since we have fixed x_0 and x_0' as the starting points, if, for example, $\eta_{11} > 0$, then we should place the corresponding edge of X_1 in a positive direction. Since $\sum_{j=1}^{k_i} \eta_{ij} = 0$, this becomes possible. More precisely, define $\lambda(\xi_i, \lambda_{j_i}, t)$ on $[\frac{1}{4}, \frac{1}{4} + \frac{1}{q+r}]$ as follows: place an edge of X_1 corresponding to positive η_{ij} in a positive direction, i.e., starting from x_0 and ending at x_1, then followed by another edge of X_1 corresponding to some negative η_{1i}. In this way, $S(\xi_i, \lambda_{j_i}, t)$ starts at x_0 and goes to x_1 and comes back to x_0. Continuing this process, we can place all edges of X_1 on $[\frac{1}{4}, \frac{1}{4} + \frac{1}{r+q} \frac{1}{2}]$. Notice that at $t = \frac{1}{4} + \frac{1}{q+r} \frac{1}{2}$, $S(\xi_i, \lambda_{j_i}, t) = x_0$. Similarly, we can place those edges of $X_2, X_3, \ldots, X_q$ on $[\frac{1}{4} + \frac{1}{q+r} \frac{1}{2}, \frac{1}{4} + \frac{q}{q+r} \frac{1}{2}]$, and still end with x_0. To realize $\eta_1, \ldots, \eta_r$, we just place those circles of X on $[\frac{1}{4} + \frac{q}{q+r} \frac{1}{2}, \frac{1}{4} + \frac{1}{2}]$, according to the signs. Notice that this is possible since a circle starts from x_0 and comes back to x_0.

For $L_1, \ldots, L_{k_1'}$, the construction would be the same. Notice that this construction can be carried out on other parts of X' as well. Hence we extend R from the vertices of X' to whole X'. We will denote the map by ϕ.

Case 2. Each block of the first row of ϕ has a zero entry.

Now $a_i \leq 0$ for $i = 0, 1, \ldots, q$. So $a_i = 0$ for all i. For convenience, we may assume that the first entry of each block of the first row of ϕ is zero. Under this assumption, $\phi_{00} = 0$. Hence $\phi_{01} = \phi_{20} = \cdots = \phi_{n_1'0} = 0$ since $a_0 = 0$. This says that we can not get a full vertex of X at x_0' and x_1'. In this case, all the $\{\lambda(\xi_i, \lambda_{j_i}, t)\}$ are the constant maps. It is easy to see that there will be no full vertex of X at $x_1', \ldots, x_{q'}'$ for ψ as well. Again, let $W(t) \in M_{n'}(C(X_1'))$ be a unitary as in Case 1, and define a map

$$R(f)(t) = W(t) \begin{bmatrix} f(\lambda(\xi_i, \lambda_{j_1}, t)) & & \\ & f(\lambda(\xi_2, \lambda_{j_2}), t) & \\ & & \ddots \end{bmatrix} W^*(t)$$

for all $f \in A$ and $t \in L_1 \vee L_2 \vee \ldots \vee L_{n_1'}$.

Clearly, the same construction extends to R to a map ψ on whole X'. This defines a map from A to B.

We now show that both the two cases give the right K_0 and K_1. To show that we did not change the K_1 map, first notice that the restriction of ψ on I, denoted by $\psi|_I$, is a $*$-homomorphism from I to J. Furthermore we have the following diagram:

$$
\begin{array}{ccc}
I & \hookrightarrow & A \\
\psi|_I \downarrow & & \downarrow \psi \\
j & \hookrightarrow & B
\end{array} \, .
$$

This induces a commutative diagram on K_1 groups

$$
\begin{array}{ccc}
K_1(I) & \xrightarrow{\imath_A} & K_1(A) \\
K_1(\psi|_I) \downarrow & & \downarrow K_1(\psi) \\
K_1(J) & \xrightarrow{\imath_B} & K_1(B)
\end{array}
$$

where $K_1(\psi|_I)$ and $K_1(\psi)$ are the corresponding maps on K_1 groups for $\psi|_I$ and ψ respectively.

Let us consider Case 1 first. Identify an edge L of X, say the first edge in the first part of X, with $[0,1]$. Let U be a unitary in A,

$$
U(t) = \left\{
\begin{array}{l}
\begin{bmatrix} e^{2\pi i t} & & & & \\ & 1 & & & \\ & & \ddots & & \\ & & & 1 \end{bmatrix} \quad t \in L \\[20pt]
\begin{bmatrix} 1 & & & \\ & 1 & & \\ & & \ddots & \\ & & & 1 \end{bmatrix} \quad t \in X \backslash L
\end{array}
\right. \, .
$$

Then $U \in I^\sim$, the ideal I with identity adjoined. So $[U] \in K_1(I^\sim) = K_1(I)$. We can identity this element in $K_1(I) = \mathbb{Z}^{(\sum\limits_{j=1}^{q} k_i + r)}$ with $(1,0,0,\ldots 0)^{tr}$. To compute $K_1(\psi|_I)([U])$, we first have the following computation:

$$
\begin{aligned}
(\psi|_I)(U-I)+I &= W(t) \begin{bmatrix} (U-1)(S(\xi_i,\lambda_{j_1},t) & & & \\ & (U-1)(S(\xi_2,\lambda_{j_2}),t) & & \\ & & \ddots & \end{bmatrix} W^*(t)+I \\[10pt]
&= W(t) \begin{bmatrix} (U-1)(S(\xi_i,\lambda_{j_1},t) & & & \\ & (U-1)(S(\xi_2,\lambda_{j_2}),t) & & \\ & & \ddots & \end{bmatrix} W^*(t)
\end{aligned}
$$

for $t \in L_1 \subset X_1'$. Recall that these $\{S(\xi_i,\lambda_{j_i},t)\}$ are coming from $\{\lambda(\xi_i,\lambda_{j_i},t)\}$. In fact, we have only changed one of them on each edge. So we may assume that we have changed $\lambda(\xi_i,\lambda_{j_i},t)$. For all other $\{S(\xi_i,\lambda_{j_i},t)\}$, they are constants or paths from one vertex of X to another. The important thing is that the maps are the same on each edge of X_1'. If we denote the first n_1' entries of the first column of η by $\eta_1^{(1)}$, then the first n_1' entries of $\left[K_1(\psi|_I)([U])\right]$ have the form

$$
\eta_1^{(1)} + (a_1 \ldots a_1)^{tr}
$$

for some integer a_1. A similar argument shows that

$$\big[K_1(\psi|_I)([U])\big] = \eta_1 + (a_1,\ldots,a_1,\ a_2,\ldots,a_2,\ldots,\ a_{q'},\ldots,a_{q'},\ b_1,\ldots,b_{r'})^{tr}$$

where η_1 is the first column of η and a_i and b_j are integers. Notice that $\{\lambda(\xi_i,\lambda_{j_i},t)\}$ are constant maps on those r' circles of X', $b_1,b_2,\ldots$ and $b_{r'}$ must be zero. Applying π_B on this element we have

$$\imath_B\big(K_1(\psi|_I)([U])\big) = \imath_B\eta_1 = \imath_B\eta\begin{bmatrix}1\\0\\\vdots\\0\end{bmatrix} = \imath_B\eta([U])$$

$$= \phi_1\big(\imath_A([U])\big).$$

On the other hand, we also have

$$\imath_B\big[K_1(\psi|_I)([U])\big] = K_1(\psi)\big(\imath_A([U])\big),$$

and so

$$\phi_1\big(\imath_A([U])\big) = K_1(\psi)(\imath_A([U])).$$

First we notice that $[U]$ is a generator for $K_1(I)$. The same argument shows that the above equation holds for all other generators. Now since π_A is onto, we conclude that $\phi_1 = K_1(\psi)$ on $K_1(A)$.

We now consider the second case. Since all the $\{\lambda(\xi_i,\lambda_{j_i},t)\}$ are constant, $K_1(\psi|_I)([U]) = 0$ for all generators of $K_1(I)$. We will show that ϕ_1 is also trivial. In fact, by our assumption, there exists $e \in E$ such that $\phi_0(e) = \phi(e) = 0$. For convenience, we may assume that the first entry of each block of e is one. Let P be the following projection in A:

$$P = \begin{bmatrix}1 & & & \\ & 0 & & \\ & & \ddots & \\ & & & 0\end{bmatrix}.$$

Then $[P] = e$. Construct a partial unitary:

$$V = \begin{cases}\begin{bmatrix}e^{2\pi it} & & & \\ & 0 & & \\ & & \ddots & \\ & & & 0\end{bmatrix} & t \in L \\[30pt] \begin{bmatrix}1 & & & \\ & 0 & & \\ & & \ddots & \\ & & & 0\end{bmatrix} & t \in X\backslash L\end{cases}$$

where L is the first edge of X_1. Clearly, the range and the domain of V are P. Since $\phi_0 \oplus \phi_1$ preserves the dimension range, the equation

$$\phi_0 \oplus \phi_1([P], [V]) = (\phi_0([P]), \phi_1([V]))$$

tells us that the element $\phi_1([V])$ is zero since $\phi_0([P]) = 0$. This says that ϕ_1 is trivial on the first summand of $K_1(I)$ in the following sense. For U constructed before, $U \in I^\sim$ and U and V represent the same element of $K_1(I)$. Since

$$\imath_A([U]) = [U] \in K_1(A)$$

we have $\imath_B \eta([U]) = \phi_1(\imath_A([U])) = \phi_1([U]) = 0$.

By constructing different V, we can show that $\phi_1([V]) = 0$ on all the generators of $K_1(I)$. This says that ϕ_1 is trivial.

We complete the proof of the theorem by showing that $K_0(\phi) = \phi_0$. Again, we have the following diagram:

$$
\begin{array}{ccc}
A & \xrightarrow{\psi} & B \\
\downarrow & & \downarrow \\
A/I & \xrightarrow{\psi_{A/I}} & B/J
\end{array} .
$$

This gives us the commutative diagram

$$
\begin{array}{ccc}
K_0(A) & \xrightarrow{K_0(\psi)} & K_0(B) \\
\downarrow & & \downarrow \\
K_0(A/I) & \xrightarrow{K_0(\psi_{A/I})} & K_0(B/J)
\end{array}
$$

where $K_0(\psi)$ and $K_0(\psi|_{A/I})$ are the induced group homomorphisms. By our construction, $K_0(\psi|_{A/I})$ is exactly ϕ. Finally, for any $e \in E$, we have the following computation:

$$K_0(\psi)(e) = K_0(\psi|_{A/I})(e) = \phi(e) = \phi_0(e) .$$

So $K_0(\psi)$ is exactly the same as ϕ_0.

Corollary.   Let A and B be two C*-algebras of finite direct sums of basic building blocks and let ϕ_0 and ϕ_1 be group homomorphisms from $K_0(A)$ to $K_0(B)$ and from $K_1(A)$ to $K_1(B)$ respectively. If $\phi_0([1_B]) = [1_B]$ and $\phi_0 \oplus \phi_1$ preserves the dimension range, then there exists a unital $*$-homomorphism ψ from A to B such that $K_*(\psi) = \phi_0 \oplus \phi_1$.

Proof: The proof is similar to the proof of the theorem if B is a single basic building block. The general case can be reduced to this case via the quotient maps.

Chapter 7.   Uniqueness

In section 6 we were able to lift a K-group map to a $*$-homomorphism between the corresponding C*-algebras. Two such liftings can be very different. In this section we will show that under certain conditions two liftings between finite direct sums of basic building blocks are approximately unitarily equivalent on a given finite subset of approximately constant elements.

7.1. We may reduce to the case that the $*$-homomorphism is from one basic building block to another basic building block. The result for the maps from a finite direct sum of basic building blocks to another finite direct sum will follow from the proof of the theorem.

Definition.   Let A and B be two basic building blocks with special spectra X and X' and with generic fibres M_n and $M_{n'}$, respectively. We shall say that a unital $*$-homomorphism ϕ from A to B has standard form, if:

(i)   On each edge L of X', identified with $I = [0,1]$, without the end points being identified, ϕ has the following expression:

$$\phi(f)(t) = U(t) \begin{bmatrix} f(S_1(t)) & & \\ & \ddots & \\ & & f(S_\alpha(t)) \end{bmatrix} U^*(t)$$

for $t \in [0,1]$ and $f \in A$. Here $U \in M_{n'}(C[0,1])$ and $\{S_i(\cdot)\}_{i=1}^\alpha \subset C(I,X)$.

(ii)   After identifying each edge of X with $[0,1]$, each of $\{S_i(t)\}_{i=1}^\alpha$ has one of the following forms:

$$\{t, \quad 1-t, \quad \text{a vertex point}\} \, .$$

For a $*$-homomorphism ϕ with standard form, we will call $\{S_i(t)\}_{i=1}^\alpha$ the eigenvalue structure of ϕ (on L).

7.2.   Let A and B be as in 7.1 and let ϕ be a unital $*$-homomorphism from A to B. If ϕ has standard form, it induces two maps ϕ_Q and ϕ_I as follows. Denote by I and J the ideals of A and B whose elements vanish at the vertices of X and X', respectively. Then the restriction of ϕ on I gives rise to a map ϕ_I from I to J. To introduce a map from A/I to

B/J, notice that the representations of ϕ at the vertices of X' can be determined by vertex points of X, so we can define ϕ_Q from A/I to B/J. Under these restrictions, we have the following two commutative diagrams:

$$
\begin{array}{ccc}
A & \xrightarrow{\phi} & B \\
\downarrow & & \downarrow \\
A/I & \xrightarrow{\phi_Q} & B/J
\end{array}
$$

and

$$
\begin{array}{ccc}
I & \longrightarrow & A \\
\phi_I \downarrow & & \downarrow \phi \\
J & \longrightarrow & B.
\end{array}
$$

Using the notation of section 6, we have the following diagram commuting at both ends:

$$
\begin{array}{ccccccccc}
0 & \to & K_0(A) & \to & K_0(A/I) & \to & K_1(I) & \to & K_1(A) & \to & 0 \\
 & & K_0(\phi) \downarrow & & K_0(\phi_Q) \downarrow & & K_1(\phi_I) \downarrow & & K_1(\phi) \downarrow \\
0 & \to & K_0(B) & \to & K_0(B/J) & \to & K_1(J) & \to & K_1(B) & \to & 0.
\end{array}
$$

In fact, it can be checked directly that the middle square is also commutative. This is because ϕ_Q and ϕ_I are induced by ϕ.

7.3. Proposition. Let A and B be two basic building blocks and let ϕ and ψ be two unital $*$-homomorphisms from A to B. Suppose that ϕ and ψ are of standard forms. For any $\varepsilon > 0$ and for any finite subset $F \subset A$, suppose that

(i) $K_0(\phi_Q) = K_0(\psi_Q)$,

(ii) ϕ and ψ have the same eigenvalue structure.

Then there exists a unitary $U \in B$ such that

$$\|U\phi(f)U^* - \psi(f)\| < \varepsilon$$

for $f \in F$.

Proof: Since $K_0(\phi_Q) = K_0(\psi_Q)$, there exists a unitary V in B/J such that

$$V\phi_Q(f)V^* = \psi_Q(f)$$

for any $f \in A/I$. (Here we use the previous notation.) In other words, at any vertex x' of the spectrum X' of B, there exists a unitary $V_{x'}$ in the fibre of B at x' such that

$$V_{x'}\phi(f)(x')V_{x'}^* = \psi(f)(x')$$

for $f \in A$.

We are going to construct a unitary U in B which extends $\{V_{x'} \mid x'$ a vertex of $X'\}$. Clearly, it is enough to work on a single edge L of X'. We identify L with $[0,1]$. Since ϕ and ψ have the same eigenvalue structure, there exists a unitary $V \in M_{n'}(C[0,1])$ such that

$$V(t)\phi(f)(t)V^*(t) = \psi(f)(t)$$

for $t \in [0,1]$ and for $f \in A$. Here, $M_{n'}$ is the generic fibre of B.

Let $t = 1$ correspond to a vertex x' of X'. We have

$$V(1)\phi(f)(1)V^*(1) = \psi(f)(1)$$
$$V_{x'}\phi(f)(1)V_{x'}^* = \psi(f)(x')$$

for all $f \in A$. This gives us

$$V^*(1)V_{x'}\phi(f)(1)V_{x'}^*V(1) = \phi(f)(1) .$$

Let σ be a small number to be specified later. Connect I and $V^*(1)V_{x'}$ on $[1 - \sigma, 1]$ in the unitary group of the commutant of $\{\phi(f)(1) \mid f \in A\}$. Extend this unitary path to $[\sigma, 1]$ by I and denote it by $X(t)$. We take σ small enough that for $f \in F$,

$$\|\phi(f)(t) - \phi(f)(1)\| < \frac{\varepsilon}{4},$$
$$\|\psi(f)(t) - \psi(f)(1)\| < \frac{\varepsilon}{4},$$
$$\|\phi(f)(s) - \phi(f)(0)\| < \frac{\varepsilon}{4},$$
$$\|\psi(f)(s) - \psi(f)(0)\| < \frac{\varepsilon}{4},$$

for $t \in [1 - \sigma, 1]$ and for $s \in [0, \sigma]$.

On $[\sigma, 1]$, denote $V(t)X(t)$ by $U(t)$. Then on $[\sigma, 1 - \sigma]$, $U(t) = V(t)$ and $U(1) = V(1)X(1) = V_{x'}$. We then have

$$\|U(t)\phi(f)(t)U^*(t) - \psi(f)(t)\| < \varepsilon$$

for $f \in F$ on $[\sigma, 1]$. Similarly, we can construct some unitary path on $[0, \sigma]$ similar to $X(t)$ so that $U(t)$ is defined on $[0, 1]$. Notice that this construction can also be carried out on each edge, which shows that there exists a unitary $U \in B$ such that

$$\|U\phi(f)(t)U^* - \psi(f)\| < \varepsilon$$

for all $f \in F$.

Corollary.   Let ϕ and ψ be two unital $*$-homomorphisms from $A = A_1 \oplus \ldots \oplus A_n$, a finite direct sum of basic building blocks, to a basic building block B. Suppose that ϕ and ψ are of standard forms. For any $\varepsilon > 0$ and for any finite subset $F \subset A$, suppose that

(i) $K_0(\phi_Q) = K_0(\psi_Q)$,

(ii) ϕ and ψ have the same eigenvalue structure.

Then there exists a unitary $U \in B$ such that

$$\|U\phi(f)U^* - \psi(f)\| < \varepsilon$$

for $f \in F$.

Proof: It is easy to see from the proof of the proposition that the same argument works in this case.

7.4. Theorem.   Let A and B be two basic building blocks with special spectra and let ϕ and ψ be two unital $*$-homomorphisms from A to B. Let $\varepsilon > 0$ and let $F \subset A$ be a finite subset which is approximately constant to within ε. Suppose that

(i) $K_*\phi = K_*\psi$, and

(ii) ϕ and ψ have the form ensured by the Perturbation Theorem 3.2 such that the variation of each generalized eigenvalue map is within $\gamma < \frac{1}{4}$ on each edge of the spectrum of B.

It follows that there exists a unitary U in B such that

$$\|U\phi(f)U^* - \psi(f)\| < 75\varepsilon$$

for all $f \in F$.

Proof: We will perturb and deform ϕ and ψ so that we can apply Proposition 7.3.

First, let us fix some notation. Denote the spectra of A and B by X and X', and denote their generic fibres by M_n and $M_{n'}$, respectively. Suppose that X has a special vertex x_0 and q other vertices $x_1 = \{x_{1i}\}_{i=1}^{n_1}, \ldots, x_q = \{x_{qi}\}_{i=1}^{n_q}$. X can be written as

$$X = X_1 \vee X_2 \vee \ldots \vee X_q \vee X_{q+1} \ldots \vee X_{q+r}$$

where each X_i corresponds to the vertex x_i for $1 \leq i \leq q$ and where the last r X_i are circles. Each X_i of the first q parts of X has some edges connecting x_0 and x_i, say k_i of them. Finally, assume that the fibre at x_{ij} is $M_{m_{ij}}$. Similarly, we have the corresponding notation for B. For example, X' has $q'+1$ vertices, etc. From now on we will identify each edge of X and X' with $[0,1]$. We will identify the special points x_0 and x_0' with 0 unless otherwise stated.

The proof will be divided into several steps.

Step 1. In this step, we will deform ϕ and ψ to $\phi^{(1)}$ and $\psi^{(1)}$ so that they have the property that each of the generalized eigenvalue maps may only achieve a vertex value at $t = 0, \frac{1}{2}, 1$ unless it is a constant vertex over $[0,1]$. In what follows, we will only work on ϕ.

Fix an edge and denote the generalized eigenvalue maps by $\{S_i(t)\}_{i=1}^\alpha$. Let us consider $S_1(t)$ first. We may assume that $S_1(t)$ achieves a vertex value of X finitely many times. This can be done, for example, by a small perturbation as in Theorem 3.1. Since $S_1(t)$ changes within γ over $[0,1]$, $S_1(t)$ can only take one vertex of X, say x_1. So we may assume that $S_1(t_i) = x_1$ for $0 \le t_1 < t_2 < \cdots < t_\ell \le 1$. Clearly, $S_1(t)$ stays in an edge of X when t varies in $[t_1, t_2]$. Identify this edge with $[0,1]$, from x_1 to x_0.

Define a homotopy H by

$$H(s,t) = \begin{cases} (1-s)S_1(t) & t \in [t_1, t_2] \\ S_1(t) & t \in [0,1]\backslash[t_1, t_2] \end{cases},$$

for $s \in [0,1]$. $H(s,t)$ is a continuous deformation of $S_1(t)$ to

$$H(1,t) = \begin{cases} 0 & t \in [t_1, t_2] \\ S_1(t) & t \in [0,1]\backslash[t_1, t_2] \end{cases}.$$

Since $F \subset A$ is approximately constant to within ε, there exists a unitary $Y \in M_n(C[0,1])$ such that YfY^* is approximately constant to within ε on $[0,1]$ for all $f \in F$. Using this Y, we may rewrite each element g of A, composed with $S_1(t)$, as follows:

$$g\big(S_1(t)\big) = \begin{cases} Y^*\big(S_1(t)\big)(YgY^*)\big(S_1(t)\big)Y\big(S_1(t)\big) & t \in [t_1, t_2] \\ g\big(S_1(t)\big) & t \in [0,1]\backslash[t_1, t_2] \end{cases}.$$

Define a new map $\phi_1^{(1)}$ from A to B. On this edge,

$$\phi_1^{(1)}(g)(t) = W(t) \begin{bmatrix} \begin{cases} Y^*\big(S_1(t)\big)(YgY^*)\big(H(1,t)\big)Y\big(S_1(t)\big) & t \in [t_1, t_2] \\ g\big(S_1(t)\big) & t \in [0,1]\backslash[t_1, t_2] \end{cases} & & \\ & \ddots & \\ & & g(s_i(t_1)) \\ & & & \ddots \\ & & & & \ddots \end{bmatrix} W^*(t).$$

On the other edges of X', define $\phi_1^{(1)}$ to be ϕ.

Notice that $Y^*\big(S_1(t)\big)Y\big(H(1,t)\big)$ is a unitary on $[0,1]$. At $t = t_1$ and $t = t_2$ its values are both I. In fact, its value outside $[t_1, t_2]$ is also I since $H(1,t) = S_1(t)$. Now $\phi_1^{(1)}$ can be expressed as follows:

$$\phi_1^{(1)}(g)(t) = W(t) \begin{bmatrix} Y^*(S_1(t)Y(H(1,t)) & & \\ & 1 & \\ & & \ddots \\ & & & 1 \end{bmatrix} \begin{bmatrix} g(H(1,t)) & \\ & \ddots \end{bmatrix}$$

$$\begin{bmatrix} Y^*(S_1(t))Y(H(1,t)) & & \\ & 1 & \\ & & \ddots \\ & & & 1 \end{bmatrix}^* W^*(t).$$

We can have similar deformations on $[t_2, t_3], \ldots, [t_{\ell-1}, t_\ell]$, on $\{S_i(t)\}_{i=1}^\alpha$ and on the other edges of X'. We still denote the resulting map by $\phi_1^{(1)}$. Denote the new eigenvalue maps by $\{\widetilde{S}_i(t)\}_{i=1}^\alpha$ for the moment. The variations are still within γ on each edge of X'. Each of them may only achieve a vertex value on a closed interval $[a, b] \subset [0, 1]$. Furthermore, $K_*\phi = K_*\phi_1^{(1)}$. Notice that if we compare $\phi_1^{(1)}$ with ϕ directly, we have

$$\|\phi_1^{(1)}(f) - \phi(f)\| < \varepsilon$$

for $f \in F$. This is because the constructions were carried out on different parts of Y and on different eigenvalue maps. Now the map $\phi_1^{(1)}$ can be expressed as

$$\psi_1^{(1)}(g)(t) = W(t)\widetilde{Y}(t) \begin{bmatrix} g(\widetilde{S}_1(t)) & & \\ & \ddots & \\ & & g(\widetilde{S}_\alpha(t)) \end{bmatrix} \widetilde{Y}^*(t)W^*(t)$$

where $\widetilde{Y}(t) = I$ at $t = 0$ and $t = 1$.

Suppose that $\widetilde{S}_1(t) = x_1$ on $[a, b] \subset [0, 1]$. If $[a, b] = [0, 1]$, we are done. Otherwise, we are going to shrink $[a, b]$ to a single point. There are two cases to be considered.

Case 1. $0 < a < b = 1$ or $0 = a < b < 1$.

Without loss of generality, we may assume the second. $\widetilde{S}_1(t)$ now takes values in an edge of X. Let U be a unitary in $M_n(C[0, 1])$ such that UfU^* is almost constant to within ε for $f \in F$. For each element $g \in A$, we can write $g(\widetilde{S}_1(t))$ as

$$g(\widetilde{S}_1(t)) = \begin{cases} g(0) & t \in [0, b] \\ g(\widetilde{S}_1(t)) & t \in [b, 1] \end{cases}$$
$$= \begin{cases} U^*(0)(UgU^*)(0)U(0) & t \in [0, b] \\ g(\widetilde{S}_1(t)) & t \in [b, 1] \end{cases}.$$

Define a new eigenvalue map $\widetilde{S}_1'(t)$ as follows:

$$\widetilde{S}_1'(t) = \begin{cases} \widetilde{S}_1(b + \frac{1-b}{b}t) & t \in [0, b] \\ \widetilde{S}(1) & t \in [b, 1] \end{cases}.$$

The following homotopy shows that $\widetilde{S}_1(t)$ and $\widetilde{S}_1'(t)$ are homotopic:

$$H(s, t) = s\widetilde{S}_1(t) + (1 - s)\widetilde{S}_1'(t).$$

Let $X(t) = \text{diag}(X_1(t), 1, 1, \ldots 1)$ where

$$X_1(t) = \begin{cases} U^*(0)U(\widetilde{S}_1'(t)) & t \in [0, b] \\ U^*(\widetilde{S}_1(t))U(\widetilde{S}_1'(t)) & t \in [b, 1] \end{cases}.$$

Then $X(t)$ is continuous and $X(0) = X(1) = I$. The following map is homotopic to $\phi_1^{(1)}$:

$$\phi_2^{(1)}(g)(t) = W(t)Y(t)X(t) \begin{bmatrix} g(\widetilde{S}_1'(t)) & & & \\ & g(\widetilde{S}_2(t)) & & \\ & & \ddots & \\ & & & g(\widetilde{S}_\alpha(t)) \end{bmatrix} X^*(t)Y^*(t)W^*(t).$$

Notice that for $f \in F$,

$$\|U^*(0)U(\widetilde{S}_1'(t))f(\widetilde{S}_1'(t))U^*(\widetilde{S}_1'(t))U(0) - f(0)\| < \varepsilon,$$

$$\|U^*(\widetilde{S}_1'(t))(UfU^*)(\widetilde{S}_1'(t))U(\widetilde{S}_1'(t)) - f(\widetilde{S}_1(t))\| < \varepsilon.$$

Hence, we have

$$\|\phi_2^{(1)}(f) - \phi_1^{(1)}(f)\| < \varepsilon$$

for all $f \in F$.

Apply similar deformations to $\{\widetilde{S}_i(t)\}_{i=2}^\alpha$ and on other edges of X'. Denote the resulting map by $\phi^{(1)}$.

Case 2. $0 < a < b < 1$.

In this case, we can shrink $[a, b]$. The deformation is similar to the deformation in Case 1. The only place that $\widetilde{S}_1'(t) = x_1$ is at t_1. We may move t_1 to $\frac{1}{2}$. This will add another ε.

In all cases, we can deform ϕ to $\phi^{(1)}$ and ψ to $\psi^{(1)}$. We do not change the eigenvalue maps as well as the representations at each vertex. The variations of the eigenvalue maps are within γ. Furthermore,

$$\|\phi(f) - \phi^{(1)}(f)\| < 3\varepsilon,$$

$$\|\psi(f) - \psi^{(1)}(f)\| < 3\varepsilon$$

for all $f \in F$.

Step 2.   If an eigenvalue map takes a vertex as its value at $t = \frac{1}{2}$, we will deform this eigenvalue map to a constant map.

In this deformation, we have to change the representations of $\phi^{(1)}$ at the vertices of X'. By Remark 3.2, this is possible.

The first deformation is to make those eigenvalue maps constant on $[\frac{3}{4}, 1]$. As in Step 2, this will induce a deformation on $\phi^{(1)}$. We will call the resulting map $\phi_1^{(2)}$. We have

$$\|\phi_1^{(2)}(f) - \phi^{(1)}(f)\| < \varepsilon$$

for all $f \in F$. Notice that all these deformations are inside the edges of X'.

Let us now deform at x_1' to make $S_1(1)$ to become a vertex of X, say x_1. In fact, we need to do this on all the edges connecting to x_1'. However, it is enough to illustrate on one edge. Let

$$H(\theta,t) = \theta t + S_1(t) - \frac{3}{4}\theta, \qquad t \in \left[\frac{3}{4},1\right], \ \theta \in [0, 4(1 - S_1(1))] .$$

We have

$$H(\theta,1) = \frac{1}{4}\theta + S_1(1) .$$

In particular, for $\theta = 4(1 - S_1(1))$, $H(\theta,1) = 1$. Here 1 is identified with x_1, just as we wanted. The deformation for the map on L is as follows:

$$\phi_2^{(2)}(f)(\theta,t) = W(t) \begin{bmatrix} V^*(S_1(t))(VfV^*)(H(\theta,t))V(S_1(t)) & & & \\ & f(S_2(t)) & & \\ & & \ddots & \\ & & & \ddots \end{bmatrix} W^*(t).$$

Here $t \in \left[\frac{3}{4},1\right]$. On $\left[0,\frac{3}{4}\right]$, we will not change anything since $H\left(\theta,\frac{3}{4}\right) = S_1\left(\frac{3}{4}\right)$.

Let us denote the final map by $\phi_{x_1}^{(2)}$, i.e., at $\theta = 4(1 - S(1))$. Then

$$\|\phi_{x_1}^{(2)}(f) - \phi_1^{(2)}(f)\| < \varepsilon$$

for all $f \in F$.

If, for example, $S_2\left(\frac{1}{2}\right)$ is a vertex, we may have to do the similar deformation. This will happen at different orthogonal blocks. After this observation, we conclude that we can deform $\phi_1^{(2)}$ to $\phi_2^{(2)}$ such that an eigenvalue map does not take a vertex as its value in $(0,1)$ unless it takes the same vertex at $t = 0$ and at $t = 1$.

Now the new $S_1(t)$ has the property that $S_1(0) = S_1\left(\frac{1}{2}\right) = S_1(1) = x_1$. The next step is to deform $S_1(t)$ to x_1 inside this edge. This is exactly the same as Step 1. We do this for each eigenvalue map and call the final map $\phi^{(2)}$. Now

$$\|\phi^{(2)}(f) - \phi^{(1)}(f)\| < 5\varepsilon .$$

Notice that the eigenvalue variations are within 2γ.

Similarly, we also have $\psi^{(2)}$ such that

$$\|\psi^{(2)}(f) - \psi^{(1)}(f)\| < 5\varepsilon$$

for all $f \in F$.

Step 3.   If an eigenvalue map achieves a vertex at one of $\{x_i'\}_{i=1}^{q'}$, we will deform it into that vertex.

Suppose that the map $\phi^{(2)}$ has the following expression on an edge of X':

$$\phi^{(2)}(f)(t) = W(t) \begin{bmatrix} f(S_1(t)) & & \\ & \ddots & \\ & & f(S_\alpha(t)) \end{bmatrix} W^*(t).$$

Assume that $S_1(1) = x_1$. There are n_1 points to form x_1. They may lie in different blocks of the fibre of B at x_1' (assume that $t = 1$ corresponds to x_1'). There is no way we can deform at this place. So we will deform the map at $t = 0$. As in Step 3, deform $S_1(0)$ to x_1 and then deform inside this edge so that the new $S_1(t)$ becomes a constant map.

Notice that if we need to deform another eigenvalue map, then it will happen in an orthogonal block. Also, if we turn to another vertex, say x_2', then we do not have to deform those previously deformed ones. They are already the vertices we need. Here we use the condition that $4\gamma < 1$. So any deformations will happen in orthogonal blocks.

The conclusion is that we are able to deform $\phi^{(2)}$ to $\phi^{(3)}$ and $\psi^{(2)}$ to $\psi^{(3)}$ in such a way that the new eigenvalue maps are constants if they achieve the vertices at $\{\xi_i'\}_{i=1}^{q'}$ and such that

$$\|\phi^{(3)}(f) - \phi^{(2)}(f)\| < 5\varepsilon,$$
$$\|\psi^{(3)}(f) - \psi^{(2)}(f)\| < 5\varepsilon$$

for all $f \in F$.

Step 4.   Deform $\phi^{(3)}$ and $\psi^{(3)}$ into standard form.

Now each eigenvalue map takes values in one edge of X. First, let us deform at x_1'. Fixing an edge of X', we have eigenvalue maps $\{S_i(t)\}_{i=1}^\alpha$. If $S_1(1)$ is not a vertex, deform it to x_0 inside an edge of X. After this deformation, the points from X corresponding to the representation at x_1' are all vertex points. We apply a similar deformation to $x_2', \ldots, x_{q'}'$. At x_0', we also deform those nonvertex points to x_0, as we did at x_1', along negative directions.

The image of each eigenvalue map now still lies in one edge of X. The next deformation is to straighten them into a linear form. Notice that if we identify all the edges with $[0, 1]$ starting from x_0 and x_0', then the only possible form is $1 - t$, if it is not a constant, on each edge of the first q' parts of X'. Here each circle of X has been identified with $[0, 1]$.

We will call the resulting map $\phi^{(4)}$ and $\psi^{(4)}$. Like Step 2, we have

$$\|\phi^{(3)}(f) - \phi^{(4)}(f)\| < 5\varepsilon,$$
$$\|\psi^{(3)}(f) - \psi^{(4)}(f)\| < 5\varepsilon$$

for all $f \in F$.

Step 5.   Deform and perturb $\phi^{(4)}$ to $\phi^{(5)}$ and $\psi^{(4)}$ to $\psi^{(5)}$ so that $K_1(\phi_I^{(5)}) = K_1(\psi_I^{(5)})$.

Since $K_*(\phi^{(4)}) = K_*(\phi) = K_*(\psi) = K_*(\psi^{(4)})$, we have the following diagram as in section 7.2:

$$0 \to K_0(A) \to K_0(A/I) \to K_1(I) \to K_1(A) \to 0$$
$$\downarrow \qquad \downarrow\downarrow \qquad \downarrow\downarrow \qquad \downarrow$$
$$0 \to K_0(B) \to K_0(B/J) \to K_1(J) \to K_1(B) \to 0 \,.$$

Here the down arrows on the two ends are $K_0(\phi)$ and $K_1(\phi)$. The two arrows from $K_0(A/I)$ to $K_0(B/J)$ are $K_0(\phi_Q^{(4)})$ and $K_0(\psi_Q^{(4)})$ and the two arrows from $K_1(I)$ to $K_1(J)$ are $K_1(\phi_I^{(4)})$ and $K_1(\psi_I^{(4)})$.

Write η for $K_1(\phi_I^{(4)}) - K_1(\psi_I^{(4)})$. Since $K_1(\phi^{(4)}) = K_1(\psi^{(4)})$, each column of η must be in the image of ∂_B. As in section 6, we may write η as follows:

$$\eta = \begin{bmatrix} \eta_{11} & \eta_{12} & \cdots \\ & \cdots & \\ \eta_{11} & \eta_{12} & \cdots \\ & \cdots & \\ 0 & \cdots & 0 \\ 0 & \cdots & 0 \end{bmatrix} \begin{array}{l} \left.\rule{0pt}{22pt}\right\} k_1' \\[18pt] \left.\rule{0pt}{14pt}\right\} r' \end{array} \,.$$

Let L be the first edge of X_1. $\eta_{11} < 0$ means that $\phi^{(4)}$ places $|\eta_{11}|$ more L on each edge of X_1' than $\psi^{(5)}$ does. So there are at least $|\eta_{11}|$ of x_0 at x_1'. Fix $|\eta_{11}|$ of them; then there are $|\eta_{11}|$ of eigenvalue maps going with them on each edge of X_1'. Deform those eigenvalue maps to the constant x_0 on $\left[\frac{1}{2}, 1\right]$. This deformation will add an ε.

If $\eta_{12} < 0$ again, it means that $\phi^{(4)}$ places $|\eta_{12}|$ more of the second edge of X_1 on each edge of X_1' than $\psi^{(4)}$ does. So at x_1', there are at least $|\eta_{11}| + |\eta_{12}|$ of x_0. Hence, we are able to choose $|\eta_{12}|$ eigenvalue maps on each edge of X_1', which achieve x_0 at x_1' and which are orthogonal to the previous ones. Deform them so that they are constant equal x_0 on $\left[\frac{1}{2}, 1\right]$. We go through all the negative entries of η and we also go through $x_1', \ldots, x_{q'}'$. The total deformation only adds an ε because we worked on different vertices of X' and different orthogonal blocks.

Deform $-\eta_{11}$ of x_0 to x_1 inside the first edge of X_1, deform $-\eta_{12}$ of x_0 to x_1 inside the second edge of X_1, etc. We also do similar deformations at $x_2', \ldots, x_{q'}'$. The resulting map $\phi_1^{(5)}$ differs by 2ε on F with $\phi^{(4)}$.

The image of each eigenvalue maps of $\phi_1^{(5)}$ may take values in two edges of X. If an eigenvalue map does lie in two edges of X, then at $t = \frac{1}{2}$, it takes x_0 as its value. We first deal with these eigenvalue maps. If the two parts lie on one edge of X, we deform this eigenvalue map to a constant vertex. If the two parts lie on two edges of X, we deform it to the form that on $\left[0, \frac{1}{2}\right]$, it is $1 - 2t$ and on $\left[\frac{1}{2}, 1\right]$, it is $2t - 1$.

Our next step is to perturb a little so that the eigenvalue maps still lie in one edge of X. Let us look at the first edge of X_1' again. We add $|\eta_{11}|$ of the first edge of X_1 going in the positive direction. But originally we have $|\eta_{11}|$ of that edge going in the negative direction. Some of them may be cancelled out during the above deformation. That happens when two lie in one edge of X. For those uncancelled ones, we fix the same number of the first edge of X_1 going in the negative direction. Deform them so that on $\left[0, \frac{1}{2}\right]$ they become x_1 and on $\left[\frac{1}{2}, 1\right]$ they become $1 - 2t$. Notice that this operation is inside each edge of X'. After we have done this on each edge of X', we get a new map $\phi_2^{(5)}$. Clearly,

$$\|\phi_1^{(5)}(f) - \phi_2^{(5)}(f)\| < \varepsilon$$

for all $f \in F$.

It is enough to examine the situation on the first edge of X_1'. Without loss of generality, we may assume that the eigenvalue maps are $\{S_i(t)\}_{i=1}^{\alpha}$, and

$$S_1(t) = \begin{cases} 1 - 2t & t \in [0, \frac{1}{2}] \\ 2t - 1 & t \in [\frac{1}{2}, 1] \end{cases}$$

and

$$S_\alpha(t) = \begin{cases} 1 & t \in [0, \frac{1}{2}] \\ 2 - 2t & t \in [\frac{1}{2}, 1] \end{cases}.$$

We may put this in the following picture:

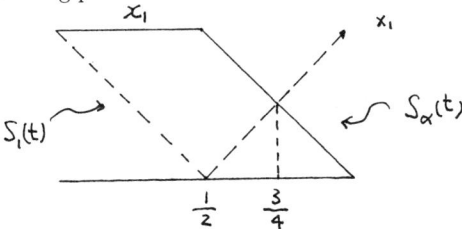

What we will do next is to perturb $\phi_1^{(5)}$ a little around $\frac{3}{4}$ so that we interchange the two eigenprojections to cancel the crossover. This is a technique used in [8].

More precisely, on this edge of X', write

$$\phi_2^{(5)}(f)(t) = W(t) \begin{bmatrix} f(S_1(t)) & & \\ & \ddots & \\ & & f(S_\alpha(t)) \end{bmatrix} W^*(t).$$

Define two new eigenvalue maps as follows:

$$\widetilde{S}_1(t) = \begin{cases} S_1(t) & t \in [0, \frac{3}{4}] \\ S_\alpha(t) & t \in [\frac{3}{4}, 1] \end{cases}$$

and

$$\widetilde{S}_\alpha(t) = \begin{cases} S_\alpha(t) & t \in [0, \frac{3}{4}] \\ S_1(t) & t \in [\frac{3}{4}, 1] \end{cases}.$$

It is easy to see that both $\widetilde{S}_1(t)$ and $\widetilde{S}_\alpha(t)$ can be deformed into eigenvalue maps such that each of them lies in a single edge of X.

Choose a small number $\sigma > 0$ to be specified later. Define a unitary path from I to the permutation unitary matrix that permutes the first and the αth diagonal entries on $\left[\frac{3}{4} - \sigma, \frac{3}{4}\right]$. Extend constantly on $\left[0, \frac{3}{4} - \sigma\right] \cup \left[\frac{3}{4}, 1\right]$. Denote this unitary path by $Z(t)$. Notice that $Z(t) \in M_{n'}(C[0, 1])$. Define a map on this edge of X' by

$$\phi_3^{(5)}(f)(t) = W(t)Z(t) \begin{bmatrix} f(\widetilde{S}_1(t)) & & & \\ & f(S_2(t)) & & \\ & & \ddots & \\ & & & f(\widetilde{S}_\alpha(t)) \end{bmatrix} Z^*(t))W^*(t).$$

Clearly,

$$\phi_3^{(5)}(f)(t) = \phi_2^{(5)}(f)(t)$$

on $\left[0, \frac{3}{4} - \sigma\right] \cup \left[\frac{3}{4}, 1\right]$ for all $f \in A$. On $\left[\frac{3}{4} - \sigma, \frac{3}{4}\right]$, $d\big(S_1(t), \widetilde{S}_\alpha(t)\big) \leq 2\sigma$. Take σ so small that for $f \in F \cup \{$generators for $K_*(A)\}$, denoted by G,

$$\|f\big(\widetilde{S}_1(t)\big) - f\big(\widetilde{S}_\alpha(t)\big)\| < \varepsilon.$$

Using the fact that $Z(t)$ commutes with

$$\begin{bmatrix} a_1 & & & & \\ & a_2 & & & \\ & & \ddots & & \\ & & & a_{\alpha-1} & \\ & & & & a_1 \end{bmatrix}$$

where each a_i has the proper size, we have

$$\|\phi_3^{(5)}(f) - \phi_2^{(5)}(f)\| < 2\varepsilon,$$

for all $f \in F \cup G$. Since $\phi_3^{(5)}$ and $\phi_2^{(5)}$ are closed on the generators of $K_*(A)$, we have $K_*(\phi_3^{(5)}) = K_*(\phi_2^{(5)})$.

Similarly, we can perturb other pairs on orthogonal blocks, and we can perturb on other edges. We will call the resulting map $\phi_3^{(5)}$ again. Now

$$\|\phi_3^{(5)}(f) - \phi_2^{(5)}(f)\| < 2\varepsilon.$$

After this small perturbation, each changed eigenvalue map can be deformed into a constant on $\left[\frac{1}{2}, 1\right]$. Hence, each new eigenvalue map lies in a single edge of X. We deform once again to put each eigenvalue map into linear form. We will call the final map $\phi^{(5)}$. It is clear that $\phi^{(5)}$ and $\phi_3^{(5)}$ differ by at most 2ε on F.

Now we can write $K_1(\phi_{4I}^{(5)}) - K_1(\psi_I^{(4)}) = \eta$ where η has non-negative entries. It is clear that similar deformation and perturbation of the eigenvalue maps of $\psi^{(4)}$ will make η zero. We will call the resulting map $\psi^{(5)}$.

Notice that each eigenvalue map of $\phi^{(5)}$ and $\psi^{(5)}$ has the form $1 - t$ unless it is a constant map, and

$$\|\phi^{(4)}(f) - \phi^{(5)}(f)\| < 12\varepsilon,$$
$$\|\psi^{(4)}(f) - \psi^{(5)}(f)\| < 12\varepsilon$$

for all $f \in F$.

Step 6. Perturb and deform $\phi^{(5)}$ and $\psi^{(5)}$ to $\phi^{(6)}$ and $\psi^{(6)}$ so that they have the same eigenvalue structure and conditions of proposition 7.3 hold.

Now the only thing missing is that they may have different constant eigenvalue maps. Let us denote the difference $K_0(\phi_Q^{(5)}) - K_0(\psi_Q^{(5)})$ by h. Then $\partial_B(h) = 0$ and $h\big(K_0(A)\big) = 0$. So each block of each row of h is the same and the sum of the $q + 1$ numbers is zero. More precisely, we may write h as follows:

$$h = \begin{bmatrix} h_{00}, h_{01} \ldots h_{01}, h_{02} \ldots h_{0q} \ldots h_{0q} \\ h_{10}, h_{11} \ldots h_{11}, h_{12} \ldots h_{1q} \ldots h_{1q} \\ \cdots \\ h_{n_1'0}, h_{n_1'1} \ldots h_{n_1'1}, h_{n_1'2} \ldots h_{n_1'q} \ldots h_{n_1'q} \\ \cdots \end{bmatrix}$$

where $\sum\limits_{j=0}^{q} h_{ij} = 0$ for each i. Using $\partial_B h = 0$, we also have the equations:

$$\sum_{j=n_1'+\cdots+n_k'+1}^{n_1'+\cdots+n_{k+1}'} h_{ji} = h_{0i}$$

for $i = 1, 2, \ldots q$ and $k = 0, 1, 2, \ldots q'$, with $n_0' = 0$.

Our goal is to make h to become zero. Without loss of generality, we may assume that $h_{01} \geq 0$, $h_{11}, \ldots, h_{\ell 1} \geq 0$ and $h_{\ell+1 1}, \ldots, h_{n_1'1} \leq 0$. So for the representations of $\phi^{(5)}$ and $\psi^{(5)}$ at x_0', $\phi^{(5)}$ has h_{01} more x_1 than $\psi^{(5)}$ has. Since the differences between the eigenvalue maps of $\phi^{(5)}$ and $\psi^{(5)}$ on each edge of X' are constant, $\phi^{(5)}$ has h_{01} more eigenvalue maps x_1 than $\psi^{(5)}$ has, on each edge of X'.

Inside the first block of fibre of B at x_1', $\phi^{(5)}$ has h_{11} more x_1 than $\psi^{(5)}$ has. Fix an edge of X_1', and suppose that the eigenvalue maps are $\{S_i(t)\}_{i=1}^\alpha$. There are two kinds of x_1 in this block. The first kind is such that the whole $S_i(1)$ representing this x_1 is in this block. The second kind are the points formed by part of $S_i(1)$. But by the perturbation theorem and by our construction, vertex points in each block do not form a full vertex at the very beginning. The later deformation only adds the full vertices. In other words, there exist h_{11} of x_1 of the first kind in the first block of the fibre of B at x_1'. A similar result holds for other blocks. Notice that if $S_i(1)$ is a full vertex x_1, then $S_i(0)$ must be also x_1 since all the non-constant eigenvalue maps have the form $1 - t$.

Since $j_{11} + \cdots + h_{\ell 1} \geq h_{01}$, there are h_{01} of x_1 at x_1' such that all of them are of the first kind. Fix h_{01} of x_1 at x_0'. At each of $\{x_i'\}_{i=1}^{q'}$, the same argument works. So we can fix h_{01} of x_1 of the first kind. To get h_{01} of x_1 over the whole of X', we need to do some perturbation at x_0'. This will be an operation on each edge of X'. Let L_1' be the first edge of X_1' and let $\{S_i(t)\}_{i=1}^{\alpha}$ be the eigenvalue maps. Without loss of generality, we may assume that $\{S_1(1), \ldots, S_{h_{01}}(1)\}$ are the h_{01} of x_1 we have chosen. As we pointed out before, all those eigenvalue maps are constant over $[0,1]$. In particular, $S_1(0), S_1(0), \ldots, S_{h_{01}}(0)$ are all x_1. These x_1 may not be the x_1 we previously fixed at x_0'. But we can interchange them as we did in Step 5.

After this observation, we conclude that, after perturbing $\phi^{(5)}$ to $\phi_1^{(6)}$, we may assume that on any edge of X', $S_i(0) = x_1$ is a vertex we fixed at x_0' if $S_i(1)$ is. This small perturbation will not change the K-group maps and $\phi^{(5)}$ and $\phi_1^{(6)}$ will differ by at most ε on F.

Deform all the fixed x_1 to x_0 via a fixed edge of X_1. Then straighten them into constant maps x_0. We will call the resulting map $\phi_2^{(6)}$. Clearly, for $f \in F$ we have

$$\|\phi_1^{(6)}(f) - \phi_2^{(6)}(f)\| < 2\varepsilon .$$

Still denote $K_0(\phi_{2Q}^{(6)}) - K_0(\psi_Q^{(5)})$ by h; then $h_{01} = 0$. The induced map $K_1(\phi_{2I}^{(6)})$ is the same as $K_1(\phi_I^{(5)})$. Furthermore, all the non-constant eigenvalue maps have the form $1 - t$.

Similarly, we can make $h_{02}, \ldots, h_{0q}$ zero. This will happen in orthogonal blocks and on $\psi^{(5)}$. Finally, we will call the resulting map $\phi_3^{(6)}$ and $\psi_3^{(6)}$. Then,

$$\|\phi_3^{(6)}(f) - \phi^{(5)}(f)\| < 3\varepsilon,$$
$$\|\psi_3^{(6)}(f) - \psi^{(5)}(f)\| < 3\varepsilon.$$

Let us still write $K_0(\phi_{3Q}^{(6)}) - K_0(\psi_{3Q}^{(6)}) = h$, where

$$h = \begin{bmatrix} 0, 0, \ldots, 0 \\ h_{10}, h_{11} \ldots h_{11} \ldots \\ \cdots \\ h_{n'_1 0}, h_{n'_1 1} \ldots h_{n'_1 1} \ldots \\ \cdots \end{bmatrix} .$$

Here h_{00} must be zero since all the other entries of the first row are zero. For convenience, we may assume that $h_{11}, \ldots, h_{\ell 1} \geq 0$ and $h_{\ell+1}, \ldots, h_{n'_1 1} \leq 0$. We can find $h_{11} + h_{21} + \cdots + h_{\ell 1}$ of x_1 for $\phi_3^{(6)}$ at x_1' and $-(h_{\ell+11} + \cdots + h_{n'_1 1})$ of x_1 for $\psi_3^{(6)}$ at x_1'. They are all of the first kind. So they are the end points of some eigenvalue maps which are necessarily constant since the eigenvalue maps all have the form $1 - t$ unless they are constants.

For $\phi_3^{(6)}$, deform these x_1 to x_0, and for $\psi_3^{(6)}$ deform the $-(h_{\ell+11} + \cdots + h_{n_1'1})$ of x_1 to x_0, both via a fixed edge of X_1, say the first edge. Secondly, straighten these maps into linear form. Denote them by $\phi_4^{(6)}$ and $\psi_4^{(6)}$. We have

$$\|\phi_4^{(6)}(f) - \phi_3^{(6)}(f)\| < 2\varepsilon$$

for all $f \in F$.

Write $K_0(\phi_{4Q}^{(6)} - K_0(\psi_{4Q}^{(6)})) = h$ again; then

$$h = \begin{bmatrix} 0 & 0 & \cdots & 0 \\ h_{01} + h_{11} & 00\ldots0h_{12} & \cdots & h_{1q} \\ h_{02} + h_{21} & 00\ldots0h_{22} & \cdots & h_{2q} \\ & & \cdots & \\ h_{n_1'0} + h_{n_1'1} & 00\ldots0h_{n_1'2} & \cdots & h_{n_1'q} \\ & & \cdots & \end{bmatrix}.$$

Here, these $h_{11} + \cdots + h_{\ell1}$ copies of x_1 deformed to x_0 are distributed as follows: h_{11} of them in the fibre of x_{11}', h_{21} of them in the fibre of $x_{12}', \ldots, h_{\ell1}$ of them in the fibre of $x_{1\ell}'$, etc.

Now $K_1(\phi_{3I}^{(6)})$ and $K_1(\psi_{3I}^{(6)})$ have been changed. We have the following:

$$(K_1(\phi_{4I}^{(6)}) = K_1(\psi_{3I}^{(6)}) + \left.\begin{bmatrix} (h_{11} + \cdots + h_{\ell1}) & 0\cdots0 \\ \vdots & \cdots \\ (h_{11} + \cdots + h_{\ell1}) & 0\cdots0 \\ 0 & \cdots \\ \vdots & \\ 0 & \cdots \end{bmatrix}\right\} k_1',$$

$$(K_1(\psi_{4I}^{(6)}) = K_1(\psi_{3I}^{(6)}) + \left.\begin{bmatrix} -(h_{\ell+11} + \cdots + h_{n_1'1}) & 0\cdots0 \\ \vdots & \cdots \\ -(h_{\ell+11} + \cdots + h_{n_1'1}) & 0\cdots0 \\ 0 & \cdots \\ \vdots & \\ 0 & \cdots \end{bmatrix}\right\} k_1'.$$

Hence, $K_1(\phi_{4I}^{(6)}) = K_1(\psi_{4I}^{(6)})$.

Similarly, we can change $x_2, \ldots x_q$ to x_0. We will denote the final maps by $\phi^{(6)}$ and $\psi^{(6)}$ with $K_1(\phi_I^{(6)}) = K_1(\psi_I^{(6)})$. Write $K_0(\phi_Q^{(6)}) - K_0(\psi_Q^{(6)}) = h$; then

$$h = \begin{bmatrix} 0 & 0\cdots0 \\ h_{10} + \cdots + h_{1q}, & 0\cdots0 \\ h_{20} + \cdots + h_{2q}, & 0\cdots0 \\ \cdots & 0\cdots0 \end{bmatrix},$$

i.e., only the first column can be nonzero. Notice that the sum of the q numbers of each row must be zero. This shows that h must be zero. We will still call the two maps $\phi^{(6)}$ and $\psi^{(6)}$. Notice that

$$\|\phi^{(6)}(f) - \phi_4^{(6)}(f)\| < 2\varepsilon,$$
$$\|\psi^{(6)}(f) - \psi_4^{(6)}(f)\| < 2\varepsilon$$

for all $f \in F$.

Step 7.   The proof of the theorem.

By Proposition 7.3, there exists a unitary $U \in B$ such that

$$\|U\phi^{(6)}(f)U^* - \psi^{(6)}(f)\| < \varepsilon$$

for all $f \in F$. Notice that on F, ϕ and $\phi^{(6)}$ differ by at most 37ε and ψ and $\psi^{(6)}$ differ by at most 37ε. Finally, we have

$$\|U\phi(f)U^* - \psi(f)\| < 75\varepsilon$$

for all $f \in F$.

Corollary.   Let $A = A_1 \oplus \cdots \oplus A_n$ and $B = B_1 \oplus \cdots \oplus B_m$ be two C*-algebras of finite direct sums of basic building blocks with special spectra. Let ϕ and ψ be two unital *-homomorphisms from A to B, and let F be a finite subset of A such that the set of components of the elements of F in each A_i is approximately constant to within a given $\varepsilon > 0$. Suppose that

(i) $K_*\phi = K_*\psi$, and

(ii) ϕ and ψ have the form ensured by Corollary 3.2 such that the variation of each
generalized eigenvalue map is at most $\gamma < \frac{1}{4}$ on each edge of the spectrum of
B_i for $i = 1, 2, \ldots, m$.

It follows that there exists a unitary U in B such that

$$\|U\phi(f)U^* - \psi(f)\| < 75\varepsilon$$

for all $f \in F$.

Proof: By passing to the quotients we may assume that $m = 1$. It is easy to see from the proof that the same argument allows us to deform and perturb ϕ and ψ so that they satisfy the conditions of Corollary 7.3. This completes the proof of the corollary.

Chapter 8.   Classification

8.1.   It was shown in 5.3 that for a unital $*$-homomorphism ϕ from a basic building block A to another basic building block B satisfying certain conditions (see Theorem 5.3), there exists a basic building block $B^0 \subset B$ and a unital $*$-homomorphism ϕ' from A to B^0 such that the following diagram commutes approximately on a given finite subset $F \subset A$ to within a given $\varepsilon > 0$:

$$
\begin{array}{ccc}
A & \xrightarrow{\phi} & B \\
\phi' \searrow & & \uparrow \\
& B^0 &
\end{array}
$$

Furthermore, on each edge of the spectrum of B° ϕ' can be expressed as

$$
\phi'(f)(t) = U(t) \begin{bmatrix} f(\tilde{\xi}_1(t)) & & \\ & \ddots & \\ & & f(\tilde{\xi}_p(t)) \end{bmatrix} U^*(t)
$$

for each $f \in A$. Recall that the variation for each of $\{\tilde{\xi}_i(t)\}_{i=1}^p$ is to within a small number $2^{a+5}\delta$ (see Theorem 5.3).

Using the argument in section 7, one can perturb ϕ' to ϕ'' so that ϕ'' has the same form as ϕ' except that those maps $\{\tilde{\xi}_i(t)\}_{i=1}^p$ have the following property: each $\tilde{\xi}_i(t)$ is not a vertex for t in the interior of $[0,1]$ unless the map is a constant one. To achieve this, one just deforms those maps, as in section 7. Of course, the variations are within $3 \cdot 2^{a+5}\delta$. Hence ϕ'' and ϕ' will be close on a given finite set if δ is small. Now each of these maps take values in an edge of the spectrum of A when t runs through $[0,1]$.

The purpose of doing this is to ensure the composition of two maps of this kind to have the same property. More precisely, let A, B and C be three C*-algebras, each of which is a finite direct sum of basic building blocks. Suppose ϕ from A to B and ψ from B to C are two unital $*$-homomorphisms with the stated property. It follows that $\psi \circ \phi$ has the same property.

Recall that we have assumed that both the C*-algebras and the connecting maps are unital.

8.2. Proposition. Let $A_1 \to A_2 \to \cdots$ be a sequence of finite direct sums of basic building blocks, and suppose that the C*-algebra inductive limit is of real rank zero. Then there exists a sequence $B_1 \to B_2 \to \cdots$ of finite direct sums of basic building blocks, with the same C*-algebra inductive limit, having the following properties:

(1) Each B_n is a finite direct sum of basic building blocks with special spectra in the sense of 5.1.

(2) The connecting maps have the forms obtained in Corollary 3.1. More precisely, let $\phi_{n,n+1}$ be the map from B_n, a finite direct sum of k copies of basic building blocks, to B_{n+1} and let ϕ be this map composed with the quotient map from B_{n+1} to a summand. Then for $f_1 \oplus f_2 \oplus \cdots \oplus f_k \in B_n$ and t in an edge L of the spectrum of that summand, ϕ can be written as:

$$\phi(f_1 \oplus \cdots \oplus f_k)(t)$$

$$= U(t) \begin{bmatrix} \begin{bmatrix} f_1(S_1(t)) & & \\ & \ddots & \\ & & f_1(S_p(t)) \end{bmatrix} & & \\ & \ddots & \\ & & \begin{bmatrix} f_k(\eta_1(t)) & & \\ & \ddots & \\ & & f_k(\eta_q(t)) \end{bmatrix} \end{bmatrix} U^*(t)$$

where $U(t)$ and $\{S_i(t)\}_{i=1}^p \cup \ldots \cup \{\eta_j(t)\}_{j=1}^q$ are continuous on L; and, furthermore, each of these maps $\{S_i(t)\}_{i=1}^p \cup \ldots \cup \{\eta_j(t)\}_{j=1}^q$ stays in an edge when t runs through L.

(3) The variations of $\{S_i(t)\}_{i=1}^p \cup \ldots \cup \{\eta_j(t)\}_{j=1}^q$ are so small that when ϕ is composed with any of $\{\phi_{1,n}, \phi_{2,n}, \ldots, \phi_{n-1,n}\}$, the corresponding eigenvalue maps have variations less than $\frac{1}{2^{n+1}}$ (see 8.1).

Proof: Recall that in the proof of Theorem 5.5, one had the following approximately commutative diagram:

$$\begin{array}{ccccccccc} A_{n_1} & \longrightarrow & A_{n_2} & \longrightarrow & A_{n_3} & \longrightarrow & A_{n_4} & \longrightarrow & \cdots \\ \uparrow & \searrow & \uparrow & \searrow & \uparrow & \searrow & \uparrow & & \\ A_1^\circ & \longrightarrow & A_2^\circ & \longrightarrow & A_3^\circ & \longrightarrow & A_4^\circ & \longrightarrow & \cdots \end{array}$$

Each eigenvalue map of the first up map does not stay in an edge. But we can overcome this by adding some vertices to the spectrum of each summand of A_{n_1}. In another word, we might assume that the points we identified in Theorem 5.3 were all vertices. It is easy to see now that we can choose n_2 large enough that (2) and (3) hold. Once n_2 is fixed, we may add vertices to the spectrum of each summand of A_{n_2}. Choose n_3 large enough so that (2) and (3) hold. Continuing this way and letting $B_n = A_n^0$, we complete the proof of the proposition.

As a consequence, one has the following corollary.

Corollary. Let $B_1 \to B_2 \to \ldots$ be a sequence of finite direct sums of basic building blocks, and suppose that the C^*-algebra inductive limit is of real rank zero. If the sequence has the three properties given in Proposition 8.1, it follows that any subsequence has these three properties. Furthermore, for any n, any finite subset $F \subset B_n$ and any $\varepsilon > 0$, there exists $m_0 > n$ such that the image of F in B_m is approximately constant to within ε for all $m > m_0$.

Proof: The first conclusion is clear. It follows from 8.1. To see the second, let us fix a finite subset $F \subset B_n$ and $\varepsilon > 0$. On an edge L of the spectrum of a summand B_m, the connecting map ϕ can be expressed as

$$\phi(f_1 \oplus \cdots \oplus f_k)(t)$$

$$= U(t) \begin{bmatrix} \begin{bmatrix} f_1(S_1(t)) & & \\ & \ddots & \\ & & f_1(S_p(t)) \end{bmatrix} & & \\ & \ddots & \\ & & \begin{bmatrix} f_k(\eta_1(t)) & & \\ & \ddots & \\ & & f_k(\eta_1(t)) \end{bmatrix} \end{bmatrix} U^*(t),$$

where the variation of each of $\{S_i(t)\}_{i=1}^p \cup \ldots \cup \{\eta_j(t)\}_{j=1}^q$ is less than $\frac{1}{2^{m+1}}$, and $f_1 \oplus \cdots \oplus f_k \in B_n$. It becomes clear that the image of F in this summand of B_m is approximately constant to within ε when m large. This completes the proof of the corollary.

8.3 Theorem. Let A and B be separable unital C^*-algebras of real rank zero. Suppose that each of A and B is the inductive limit of a sequence of finite direct sums of basic building blocks with unital connecting maps. Suppose that the graded groups $K_*(A)$ and $K_*(B)$ are isomorphic, in a way preserving the graded dimension range. Then A and B are isomorphic.

Proof: By Proposition 8.3 and Corollary 8.3, A and B may be expressed as the inductive limits of sequences $\{A_i\}$ and $\{B_i\}$ of finite direct sums of basic building blocks, with the properties stated in Proposition 8.3 and Corollary 8.3.

The proof now follows very closely the proof for the special case of circle algebras [11]. Let us divide it into several steps.

Step 1. The isomorphism $K_*(A) \to K_*(B)$ can be lifted to an intertwining of subsequences of the sequences $\{K_*(A_i)\}$ and $\{K_*(B_i)\}$. More precisely, after passing to subsequences of $\{A_i\}$ and $\{B_i\}$ and changing notation, one has a commutative diagram

$$\begin{array}{ccccccc} K_*(A_1) & \longrightarrow & K_*(A_2) & \longrightarrow & \cdots & \longrightarrow & K_*(A) \\ \downarrow & \nearrow & \downarrow & \nearrow & & & \uparrow\downarrow \\ K_*(B_1) & \longrightarrow & K_*(B_2) & \longrightarrow & \cdots & \longrightarrow & K_*(B) \end{array}$$

where each map preserves both the graded group structure and the graded dimension range.

This is due to the special nature of A_i and B_i. We must show that the map $K_*(A_i) \to K_*(B)$ obtained by composing $K_*(A_1) \to K_*(A)$ with $K_*(A) \to K_*(B)$ can be lifted to a graded group map $K_*(A_1) \to K_*(B_{n_1})$ for some n_1, preserving the graded dimension range. Since A_1 is a sum of finitely many basic building blocks, we may suppose that A_1 is itself a basic building block. We have seen in section 6 that the positive cone of $K_0(A_1)$ is finitely generated and those generators also generate $K_0(A_1)$. We have also seen that $K_1(A)$ is a finite direct sum of cyclic groups that can be identified with $\mathbb{Z}$. This says that the lifting of the group maps $K_0(A_1) \to K_0(B)$ and $K_1(A_1) \to K_1(B)$ is then possible because $K_*(B)$ is the inductive limit of $K_*(B_n)$.

Furthermore, the lifting $K_*(A_1) \to K_*(B_{n_1})$ must preserve the graded dimension range, at least after it is composed with the map $K_*(B_{n_1}) \to K_*(B_{n_1'})$, for n_1' sufficiently large. More specifically, let $[e_1], \cdots, [e_k]$ be the generators for $K_0^+(A_1)$ and let $g_1, \cdots, g_m$ be the generators for $K_1(A_1)$. Then the nonzero elements of the graded demension range of A_1 are the pairs $([e_i], ng_j)$ for $n \in \mathbb{Z}$. We may certainly pass to the case that the finitely many elements $([e_i], g_j)$ map into the graded dimension range of B_{n_1}, and it remains to show that if an element $([e], h)$ of $K_*(B_{n_1})$ belongs to the graded dimension range of B_{n_1}, then so does $([e], nh)$ for any $n \in \mathbb{Z}$. In fact, we have seen in section 6 that $K_1(A_1)$ is a quotient of $K_1(I)$ where I is the ideal of A_1 vanishing at all vertices of the spectrum of A_1. Lift h to $\tilde{h}$ in $K_1(A)$, and let u be a unitary in I such that $[u] = \tilde{h}_1$. We may choose u in such a way that on each edge of the spectrum of A_1 u can be expressed as

$$u(t) = \begin{bmatrix} e^{2\pi i k t} I_{[e]} & \\ & I \end{bmatrix}$$

for some integer k depending on the edge. It is easy to see that if we take u_1 to be the unitary in A_1 such that on the same edge u_1 can be expressed as

$$u_1(t) = \begin{bmatrix} e^{2\pi i k n t} I_{[e]} & \\ & I \end{bmatrix}$$

then $[u_1] = nh$ and $([e], nh)$ is in the dimension range.

Repeating this procedure, first to find a map $K_*(B_{n_1}) \to K_*(A_{j_2})$, and then $K_*(A_{j_2}) \to K_*(B_{n_2})$, and so on, yields an intertwining of subsequences with the desired properties.

Step 2.   After passing to a suitable subsequence of $\{A_i\}$ and $\{B_i\}$ (and changing notation), the maps intertwining the sequences $\{K_*(A_i)\}$ and $\{K_*(B_i)\}$ can be lifted, individually, to C*-algebra homomorphisms. In other words, one has a not necessarily commutative diagram

$$\begin{array}{ccccccc}
A_1 & \longrightarrow & A_2 & \longrightarrow & \cdots & \longrightarrow & A \\
\downarrow & \nearrow & \downarrow & \nearrow & & & \\
B_1 & \longrightarrow & B_2 & \longrightarrow & \cdots & \longrightarrow & B,
\end{array}$$

Step 3. After passing again to suitable subsequences of $\{A_i\}$ and $\{B_i\}$ (and changing notation), it is possible to perturb each of the homomorphisms $A_i \to B_i$ and $B_i \to A_{i+1}$ obtained in Step 2 by an inner automorphism, in such a way that the diagram becomes an approximate intertwining, in the sense of Theorem 4.2. Hence A and B are isomorphic.

This follows from Theorem 7.4. Let $a_i = \{a_{i,n}\}$, $b_i = \{b_{i,n}\}$ be dense sequences in A_i and B_i, $i = 1, 2, \ldots$, as in Theorem 4.2. We wish to pass to subsequences of $\{A_i\}$ and $B_i\}$, and change each map $A_i \to B_i$ and $B_i \to A_{i+1}$ by composing it with an inner automorphism, in such a way as to fulfil the hypothesis of Theorem 4.2 concerning the finite sets $S_i \subset A_i$ and $T_i \subset B_i$ defined (as in Theorem 4.2) with respect to the new sequences $\{A_i\}$ and $\{B_i\}$ (subsequences of the previous ones), and the new maps $A_i \to B_i$ and $B_i \to A_{i+1}$ (inner perturbations of the previous ones). This is done by a step-by-step procedure, using the choice of sequences $\{A_i\}$ and $\{B_i\}$ made at the beginning of the proof, together with Theorem 7.4.

Specifically, first replace B_1 by B_{j_1} with j_1 sufficiently large that, after changing notation, the new set $T_1 \subset B_1$ (the image of the original one) is approximately constant (on each basic building block) to with $\frac{2^{-1}}{2}$. At the same time, replace A_2 by A_{j_2+1}, so that B_1 still maps into A_2. Second, replace A_2 by A_{i_2} with i_1 sufficiently large that, after changing notation, the new set $S_2 \subseteq A_2$ (the image of the original one) is approximately constant (on each basic building block) to within $\frac{2^{-2}}{2}$. At the same time, replace B_2 by B_{i_2}, so that A_2 still maps into B_2. Third, replace B_2 by B_{j_2} with j_2 sufficiently large that the conditions for Theorem 7.4 are satisfied. More precisely, we may assume i_2 sufficiently large that a very small perturbation of the map $B_1 \to A_{i_2}$ will make it to have the property (2) of Proposition 8.2. Furthermore, property (3) of Proposition 8.2 holds for a very small number. Similarly, take j_2 large enough that the map from a new A_2 to B_{j_2} has the above property. Now, it is possible to do the following fourth procedure. Using Theorem 7.4, applied to $T_1 \subseteq B_1$ and the two maps from B_1 to B_2 (the new B_2) – across and up and down – modify the map $A_2 \to B_2$ by an inner automorphism so that the two maps from B_1 to B_2 agree within 2^{-1} on T_1.

Fifth, and only after modifying the map $A_2 \to B_2$ as above, replace B_2 again by $B_{j_2'}$ with j_2' sufficiently large that, after changing notation, the new set $T_1 \subseteq B_2$ (the image of the original one) is approximately constant (on each basic building block) to within $\frac{2^{-1}}{2}$. At the same time, replace A_3 by $A_{j_3'}+1$, so that B_2 still maps into A_3. Sixth, replace A_3 by A_{i_3} with i_3 sufficiently large that Theorem 7.4 can be applied. Seventh, using Theorem 7.4, applied to $S_2 \subseteq A_2$ and the two maps from A_2 to A_3 (the new A_3) – across, and down and up –, modify the map $B_3 \to A_3$ by an inner automorphism so that the two maps from A_2 to A_3 agree by at most 2^{-2}.

Continuing in this way (in groups of three steps), one eventually passes to subsequences of $\{A_i\}$ and $\{B_i\}$ and modified maps $A_i \to B_i$ and $B_i \to A_{i+1}$ such that the hypotheses of

Theorem 4.2 are fulfilled. Then by Theorem 4.2 the hypotheses of Theorem 4.1 is fulfilled, and so A is isomorphic to B.

Chapter 9. Applications

9.1 The following result was obtained in [6].

Let A_θ be the irrational rotation C*-algebra generated by two unitaries U and V satisfying the relation $UV = e^{2\pi i\theta}VU$. The automorphism of order two α sending U to U^{-1} and V to V^{-1} is called the flip. It was shown in [6] that the crossed product $A_\theta \rtimes_\alpha \mathbb{Z}_2$ and fixed point sub-C*-algebra A_θ^α are AF-algebras. In fact, $A_\theta \rtimes_\alpha \mathbb{Z}_2$ was expressed as an inductive limit of matrix algebras over intervals with multiple end points. It was shown then that the algebra was of real rank zero and $K_1(A_\theta \rtimes_\alpha \mathbb{Z}_2)$ was trivial. As a consequence of Theorem 8.3, $A_\theta \rtimes_\alpha \mathbb{Z}_2$ is AF.

There are some other similar applications. For example, it was shown in [13] that a UHF algebra crossed product with $\mathbb{Z}_2$ may not be real rank zero.

9.2. In [12], it was shown that the ordered group K_*A has the Riesz decomposition property if A is a C*-algebra of real rank zero and stable rank one. As a consequence this is true for all the C*-algebras we have considered here. Next, we show that for the C*-algebras considered in this paper, one can actually replace arbitrary non-Hausdorff graphs by unit circles. In fact, by [12] there exists a real rank zero C*-algebra B that can be expressed as an inductive limit of finite direct sums of basic building blocks with spectrum the circle such that $K_*(B) \cong K_*(A)$. By Theorem 8.3, B is isomorphic to A.

9.3. In what follows we will extend Theorem 8.3 to the inductive limits of finite direct sums of matrix algebras over compact one-dimensional spaces defined as inverse limits of finite graphs. This will be a direct consequence of Theorem 9.4 and Theorem 8.3. For simplicity, we will assume in 9.3 and 9.4 that both the C*-algebras and the connecting maps in the inductive limit systems are unital.

As in [4], we have the following

Definition. If Ω is a compact Hausdorff space, we shall say that Ω has dimension $\leq n$ if Ω is the inverse limit of a sequence (Ω_m) of CW-complexes of dimension at most n or equivalently if $C(\Omega)$ is the inductive limit of a sequence of algebras $\big(C(\Omega_m)\big)$, where each Ω_m is a CW-complex of dimension at most n.

What we are interested in here are the one-dimensional spaces. Notice that in order to be compact Ω_m is necessarily finite. Denoting the metric on Ω_m by d_m, one can define a metric d on Ω. More precisely, given the following inverse limit $\Omega = \varprojlim (\Omega_m, \phi_{m,n})$, Ω can be viewed as a subspace of $\prod_{m=1}^{\infty} \Omega_m$:

$$\Omega = \{(x_m) \mid x_n = \phi_{m,n}(x_m), \quad x_n \in \Omega_n,\ x_m \in \Omega_m\}.$$

The topology is the subspace topology. It is clear that this topology is equivalent to the topology given by the following metric d:

$$d(\{x_m\}, \{y_m\}) = \sum_{m=1}^{\infty} 2^{-m} \frac{d_m(x_m, y_m)}{1 + d_m(x_m, y_m)}.$$

For each m there exists a projection $\phi_{\infty,m}$ from Ω to Ω_m. It is easy to see that we may assume that $\phi_{\infty,m}(\Omega)$ intersects each connected component of Ω. Hence, at least one point in each component of Ω_m has preimage in Ω.

Proposition. *Let X be a finite connected graph and let Ω be an one-dimensional space defined as the inverse limit of a sequence of finite graphs (Ω_m). Then for any finite subset $F \subset M_p\big(C(X)\big)$, any $\varepsilon > 0$ and any unital $*$-homomorphism ϕ from $M_p\big(C(X)\big)$ to $M_q\big(C(\Omega)\big)$, there exists an integer m_0 and a unital $*$-homomorphism ψ from $M_p\big(C(X)\big)$ to $M_q\big(C(\Omega_{m_0})\big)$ such that the following diagram commutes on F to within ε:*

$$
\begin{array}{c}
M_q\big(C(\Omega_1)\big) \\
\downarrow \\
\vdots \\
\downarrow \\
M_q\big(C(\Omega_{m_0})\big) \\
\quad\quad\psi \nearrow \quad\quad \downarrow \\
\vdots \\
\downarrow \\
M_p\big(C(X)\big) \xrightarrow{\ \phi\ } M_q\big(C(\Omega)\big).
\end{array}
$$

Here p and q are two integers.

Proof: Let d_m be the metric on Ω_m and let d be the metric on Ω introduced in the beginning of this section. Denote by $\phi_{m,n}$ the connecting map from Ω_m to Ω_n and denote by ϕ_m the map from Ω to Ω_m.

The proof is similar to the proof of Theorem 3.2. Recall that in the proof of Theorem 3.2 we introduced three numbers $\gamma > 0$, $\delta > 0$ and $\beta > 0$ and we also introduced three integers $a < R < M < N$. For M and R, form the test functions associated with them, in the sense of

2.6. There exists a $2M/N$ dense finite subset $H \cup \widetilde{H} \in M_p(C(X))$. Since Ω is compact, there exists $\sigma > 0$ such that for any w_1, w_2 in Ω with $d(w_1, w_2) < \sigma$, $\|\phi(f)(w_1) - \phi(f)(w_2)\|$ is less than γ for $f \in F$, and $\|\phi(h)(w_1) - \phi(h)(w_2)\|$ is less than δ for any test function h.

Let m_0 be an integer such that $\sum\limits_{m=m_0}^{\infty} 2^{-m} < \sigma/4$. Partition each edge of Ω_{m_0} into small intervals so that for any two points s and t in the same interval, the following inequality holds:

$$\sum_{k=1}^{m_0} 2^{-k} \frac{d_k\big(\phi_{m_0,k}(s), \phi_{m_0,k}(t)\big)}{1 + d_k\big(\phi_{m_0.k}(s), \phi_{m_1,k}(t)\big)} < \sigma/4 .$$

Assume that t_i is one partition point and it is in $\phi_{m_0}(\Omega)$. The preimages of t_i in Ω may not be unique but they are within $\sigma/4$. Pick one of them, say w_i, and define a map ψ at t_i as follows:

$$\psi(g)(t_i) \;=\; \phi(g)(w_i)$$

for $g \in M_p\big(C(X)\big)$. ψ is hence defined on some of the partition points.

Let $[t_i, t_{i+1}]$ be one of the small intervals and assume that one of the end points, say t_{i+1}, is not in $\phi_{m_0}(\Omega)$. Since Ω is compact, its image in Ω_{m_0} is closed. So the following number

$$t_0 \;=\; \sup\{t \in [t_i, t_{i+1}] \mid t \in \phi_{m_0}(\Omega)\}$$

is in $\phi_{m_0}(\Omega)$. We refine the partition of Ω_{m_0} by adding the points of this kind.

After this modification we may assume that those small intervals can be divided into two types. The first type consists of the intervals that both of the ends of each interval are in $\phi_{m_0}(\Omega)$. The second type consists of the intervals that at least one end point of each interval is not in $\phi_{m_0}(\Omega)$ and the interior of each interval does not intersect $\phi_{m_0}(\Omega)$.

Only after this observation we assign a representation at each partition point that does not belong to $\phi_{m_0}(\Omega)$, say ϕ. For example, we just define $\psi(g)$ on those points to be

$$u \begin{bmatrix} g(x_1) & & \\ & \ddots & \\ & & g(x_k) \end{bmatrix} u^*$$

where $g \in M_p\big(C(X)\big)$, $k = q/p$ and u is a unitary in $M_q(\mathbb{C})$.

Next, we are going to interpolate ψ in each interval. For the intervals of the second type, one connects them in any continuous way. Notice that this will not affect the almost commuting diagram we are going to prove. To interpolate ψ in a first type interval $[t_i, t_{i+1}]$, one can do exactly the same thing as in the proof of Theorem 3.2. More precisely, ψ was defined at t_i and t_{i+1}. Furthermore, $\|\psi(f)(t_i) - \psi(f)(t_{i+1})\|$ is less that γ for all $f \in F$ (one needs to enlarge F by adding all the matrix units of M_p), and $\|\psi(h)(t_i) - \psi(h)(t_{i+1})\|$ is less than δ for all the test functions associated with R and M. With those conditions, one can construct ψ on

$[t_i, t_{i+1}]$ such that the variation of each $\psi(f)$ is less than $\varepsilon/4$ for $f \in F$. (Here δ, γ, R and M are determined by F and ε. See Theorem 3.2.)

It remains to show that the diagram

$$
\begin{array}{ccc}
 & & M_q\big(C(\Omega_{m_0})\big) \\
 & \overset{\psi}{\nearrow} & \downarrow \\
M_p\big(C(X)\big) & \overset{\phi}{\longrightarrow} & M_q\big(C(\Omega)\big)
\end{array}
$$

approximately commutes on F to within ε.

Taking an $f \in F$, one has two elements $\phi(f)$ and $\psi(f) \circ \phi_{m_0}$ in $M_q\big(C(\Omega)\big)$. We need to prove that the difference $\|\phi(f)(w) - \psi(f) \circ \phi_{m_0}(w)\|$ is less than ε for all $w \in \Omega$. We may assume that $\phi_{m_0}(w) \in [t_i, t_{i+1}]$ is of the first type. Let $w_i \in \Omega$ such that $\phi_{m_0}(w_i) = t_i$ and such that $\psi(f))(t_i) = \phi(f)(w_i)$. Since t_i and $\phi_{m_0}(w)$ are in the same interval, $d(w_i, w) < \sigma/4$. This says that $\|\phi(f)(w_i) - \phi(f)(w)\|$ is less than δ, which can be taken to be smaller than $\varepsilon/4$. Hence

$$
\begin{aligned}
\|\psi(f)\big(\phi_{m_0}(w)\big) &- \phi(f)(w)\| \\
&\leq \|\psi(f)\big(\phi_{m_0}(w)\big) - \psi(f)(t_i)\| + \|\phi(f)(t_i) - \phi(f)(w)\| \\
&= \|\psi(f)\big(\phi_{m_0}(w)\big) - \phi(f)(t_i)\| + \|\phi(f)(w_i) - \phi(f)(w)\| \\
&< \frac{\varepsilon}{4} + \frac{\varepsilon}{4} < \varepsilon .
\end{aligned}
$$

This completes the proof of the proposition.

When the graphs are restricted to certain special forms, one may allow the matrix order q to change along the way and get a general lifting result (see [22]).

Corollary.   Let $X_1, X_2, \ldots, X_a$ be a finite connected graphs and let $\Omega_1, \Omega_2, \ldots, \Omega_b$ be b compact one-dimensional spaces, where each Ω_k is the inverse limit of a sequence of finite graphs $\{\Omega_n^k\}_{n=1}^\infty$. Suppose that ϕ is a unital $*$-homomorphism from $\bigoplus_{i=1}^{a} M_{p_i}\big(C(X_i)\big)$ to $\bigoplus_{i=1}^{b} M_{q_i}\big(C(\Omega_i)\big)$ where $\{p_i\}_{i=1}^a \cup \{q_i\}_{i=1}^b$ are positive integers. It follows that for any finite set $F \subset \bigoplus_{i=1}^{a} M_{p_i}\big(C(X_i)\big)$ and any $\varepsilon > 0$, there exists an m_0 and a unital $*$-homomorphism ψ from $\bigoplus_{i=1}^{a} M_{p_i}\big(C(X_i)\big)$ to $\bigoplus_{i=1}^{b} M_{q_i}\big(C(\Omega_{m_0}^{(i)})\big)$ such that the following diagram commutes on F to within ε:

$$
\begin{array}{ccc}
 & & \bigoplus_{i=1}^{b} M_{q_i}\big(C(\Omega_{m_0}^{(i)})\big) \\
 & \overset{\psi}{\nearrow} & \downarrow \\
\bigoplus_{i=1}^{a} M_{p_i}\big(C(X_i)\big) & \overset{\phi}{\longrightarrow} & \bigoplus_{i=1}^{n} M_{q_i}\big(C(\Omega_i)\big)
\end{array}
$$

Proof: It is easy to see that the proof for the proposition works for the case $a > 1$ and $b = 1$. But the general case can be reduced to the case of $b = 1$ by passing to the quotients.

9.4 Theorem. *Let A be a real rank zero C^*-algebra that can be expressed as an inductive limit of a sequence $\{A^{(i)}\}_{i=1}^{\infty}$, where each $A^{(i)}$ is a finite direct sum of matrix algebras over compact one-dimensional spaces. Then A can be expressed as an inductive limit of finite direct sums of matrix algebras over finite graphs.*

Proof: For each i, we may assume that there exists a sequence $\{A_n^{(i)}\}_{n=1}^{\infty}$, with each $A_n^{(i)}$ a finite direct sum of matrix algebras over finite graphs, such that $A^{(i)} = \varinjlim A_n^{(i)}$, in the sense of 9.3. We then have the following diagram:

$$
\begin{array}{ccccccc}
A_1^{(1)} & & A_1^{(2)} & & & & \\
\downarrow & & \downarrow & & & & \\
A_2^{(1)} & & A_2^{(2)} & & \vdots & & \\
\downarrow & & \downarrow & & & & \\
\vdots & & \vdots & & & & \\
\downarrow & & \downarrow & & & & \\
A^{(1)} & \longrightarrow & A^{(2)} & \longrightarrow & \cdots & \longrightarrow & A.
\end{array}
$$

The connecting map from $A^{(i)}$ to $A^{(j)}$ will be denoted by ϕ_{ij} and the map from $A_k^{(i)}$ to $A^{(i)}$ will be denoted by $\pi_k^{(i)}$. Notice that all the maps here are unital.

For each n, let $f_n = \{f_{n,k}\}_{k=1}^{\infty} \subset A^{(n)}$ be a dense subsequence. Clearly, there exists $g_1^{(1)} \in A_{n_1}^{(1)}$ for some n_1 such that the difference $\|f_{1,1} - \pi_{n_1}^{(1)}(g_1^{(1)})\|$ is less than $\frac{1}{2}$. Since $A_{n_1}^{(1)}$ is separable, there exists a dense subset $a_1 = \{a_{1,n}\}_{n=1}^{\infty} \subset A_{n_1}^{(1)}$. Fix $\sigma > 0$ to be specified later. For $M = 1$, form a σ dense finite subset of all test functions in $A_{n_1}^{(1)}$ associated with M, say H_1. Let F_1 be a subset of $A_{n_1}^{(1)}$:

$$
F_1 = \{g_1^{(1)}, a_{1,1}\} \cup H_1 .
$$

Since A is of real rank zero, there exists $k_2 > 1$ such that the eigenvalue variation of $\phi_{1,k_2} \circ \pi_{n_1}^{(1)}(h)$ is less than σ for all $h \in H_1$. By Corollary 9.3, there exists $n_2 > n_1$ and a $*$-homomorphism ψ_1 form $A_{n_1}^{(1)}$ to $A_{n_2}^{(k_2)}$ such that the following inequality holds for all $f \in F_1$:

$$
\|\pi_{n_2}^{(k_2)} \circ \psi_1(f) - \phi_{1,k_2} \circ \pi_{n_1}^{(1)}(f)\| < \sigma .
$$

Recall that in the proof of Proposition 9.3, we partitioned each edge of the spectrum of $A_{n_2}^{(k_2)}$ into small intervals. For the interval $[t_i, t_{i+1}]$ of the first type (see Proposition 9.3), the representations of ψ_1 at t_i and t_{i+1} are exactly the representations of $\phi_{1,k_2} \circ \pi_{n_1}^{(1)}$ at some preimages of w_i and w_{i+1}. Since the eigenvalue variation of $\phi_{i,k_2} \circ \pi_{n_1}^{(1)}(h)$ is less than 2σ for all the test functions h associated with M, the points back in the spectrum of $A_{n_1}^{(1)}$ can be paired to be within 6σ one by one (Lemma 2.3). For all intervals $[t_i, t_{i+1}]$ of the second type, we assigned any representation to ψ_1 at those ends that do not have preimage in the limit spaces. We may require that the points in the spectra of $A_{n_1}^{(1)}$ associated to this representation actually come from the representation of ψ_1 at an end point of a first type interval. It is easy to see now that for any two points in the spectrum of a summand of $A_{n_2}^{(k_2)}$, the points in the spectra of $A_{n_1}^{(1)}$ corresponding to the representations of ψ_1 at these two points can be paired to be within $3 \times 6\sigma = 18\sigma$ one by one. If σ is small, we may assume that

(1)   any two groups of points corresponding to ψ_1 at two points of the spectrum of a summand of $A_{n_2}^{(k_2)}$ can be paired to within $\frac{1}{2}$, and

(2) $\|\phi_{1,k_2} \circ \pi_{n_1}^{(1)}(f) - \pi_{n_2}^{(k_2)} \circ \psi_1(f)\| < \frac{1}{2}$ for all $f \in F_1$.

Similarly, let $F_2 \subset A_{n_2}^{(k_2)}$ be the finite subset consisting of $\psi_1(a_{1,1})$, $\psi_1(a_{1,2})$, $a_{2,1}, a_{2,2}$ and four elements $g_1^{(2)}, g_2^{(2)}, g_3^{(2)}$ and $g_4^{(2)}$ such that

$$\|\pi_{n_2}^{(k_2)}(g_i^{(2)}) - \phi_{1k_2}(f_{1,i})\| < \frac{1}{2^2} \quad i = 1, 2,$$

$$\|\pi_{n_2}^{(k_2)}(g_{i+2}^{(2)}) - f_{2,i}\| < \frac{1}{2^2} \qquad i = 1, 2.$$

(Here we may need to enlarge n_2 at the beginning.) Then there exist $k_3 > k_2$ and $n_3 > n_2$, and there exists ψ_2 from $A_{n_2}^{(k_2)}$ to $A_{n_3}^{(k_3)}$ such that

(1)   any two groups of points corresponding to ψ_2 at two points of the spectrum of a summand of $A_{n_3}^{(k_3)}$ can be paired to be within $\frac{1}{2^2}$, and

(2) $\|\phi_{k_2,k_3} \circ \pi_{n_2}^{(k_2)}(f) - \pi_{n_3}^{(k_3)} \circ \psi_2(f)\| < \frac{1}{2^2}$ for all $f \in F_2$.

Continuing this way, we obtain two sequences $\{k_i\}_{i=1}^{\infty}$ ($k_1 = 1$) and $\{n_i\}_{i=1}^{\infty}$ ($n_1 = 1$) and a sequence of $*$-homomorphisms $\{\psi_n\}_{i=1}^{\infty}$ as in the picture

$$
\begin{array}{ccccccccc}
A_{n_1}^{(k_1)} & \xrightarrow{\psi_1} & A_{n_2}^{(k_2)} & \xrightarrow{\psi_2} & A_{n_3}^{(k_3)} & \xrightarrow{\psi_3} & \cdots & \longrightarrow & B \\
\downarrow & & \downarrow & & \downarrow & & & & \\
A^{(k_1)} & \longrightarrow & A^{(k_2)} & \longrightarrow & A^{(k_3)} & \longrightarrow & \cdots & \longrightarrow & A,
\end{array}
$$

such that on the mth square the following holds:

$$\|\phi_{k_m,k_{m+1}} \circ \pi_{n_m}^{(k_m)}(f) - \pi_{n_{m+1}}^{(k_{m+1})} \circ \psi_m(f)\| < \frac{1}{2^m}$$

for all $f \in F_m$, where F_m consists of the images of the first m terms of $a_1, a_2, \ldots, a_{m-1}$, the first m terms of a_m and a set of elements that approximates the images of the first m terms of $f_1, f_2, \ldots, f_{m-1}$ and the first m terms of f_m, to within $\frac{1}{2^m}$, and where B is the C*-algebra inductive limit.

We complete the proof by showing that B is isomorphic to A. First, we define a map ϕ from a dense set of B to A. Let $a \in a_{n'}$, say $a = a_{n',\ell}$. There exists an integer r such that the image of $a_{n',\ell}$ in $A_{n'+r}^{(k_{n'+r})}$ belongs to $F_{n'+r}$. Denote this image by a again. We can map a into A as follows:

$$
\begin{array}{ccc}
a \in A_{n'+r}^{(k_{n'+r})} & \longrightarrow & A_{n'+r+m}^{(k_{n'+r+m})} \\
& & \downarrow \\
& & A^{(k_{n'+r+m})} \longrightarrow A,
\end{array}
$$

i.e., a is sent to $\phi_{k_{n'+r+m},\infty} \circ \pi_{n_{n'+r+m}}^{(k_{n'+r+m})} \circ \psi_{n'+r,n'+r+m}(a)$, where $\phi_{i,\infty}$ is the map from $A^{(i)}$ to A and ψ_{ij} is the map from $A_{n_i}^{(k_i)}$ to $A_{n_j}^{(k_j)}$. Let $\psi_{j,\infty}$ be the map from $A_{n_j}^{(k_j)}$ to B. We define $\phi_{n'}\big(\psi_{n'+r,\infty}(a)\big)$ in A to be the following limit

$$\lim_{m\to\infty} \phi_{k_{n'+r+m},\infty} \circ \pi_{n_{n'+r+m}}^{(k_{n'+r+m})} \circ \psi_{n'+r,n'+r+m}(a) .$$

The following computation shows that the above limit does exist. Consider

$$
\begin{array}{ccccccc}
A_{n_{n'+r}}^{(k_{n'+r})} & \longrightarrow & A_{n_{n'+r+m}}^{(k_{n'+r+m})} & \longrightarrow & A_{n_{n'+r+m+1}}^{(k_{n'+r+m+1})} & \to \cdots \to & A_{n_{n'+r+m+m'}}^{(k_{n'+r+m+m'})} \\
 & & \downarrow & & \downarrow & & \downarrow \\
 & & A^{(k_{n'+r+m})} & \longrightarrow & A^{(k_{n'+r+m+1})} & \to \cdots \to & A^{(k_{n'+r+m+m'})} & \longrightarrow & A.
\end{array}
$$

Compute

$$
\Big\| \phi_{k_{n'+r+m+m'},\infty} \circ \pi_{n_{n'+r+m+m'}}^{(k_{n'+r+m+m'})} \circ \psi_{n'+r,n'+r+m+m'}(a)
$$
$$
- \phi_{k_{n'+r+m},\infty} \circ \pi_{n_{n'+r+m}}^{(k_{n'+r+m})} \circ \psi_{n'+r,n'+r+m}(a) \Big\|
$$
$$
\leq \sum_{\ell=0}^{\infty} 2^{-(n'+r+m+\ell)} .
$$

This says that the limit does exist. Since $a_{n'}$ is dense, we can actually define a map from $A_{n'}$ to B. It follows from 2.3 of [11] that there is a $*$-homomorphism ϕ from A to B.

It remains to show that ϕ is injective and surjective. By our construction, it is easy to see that ϕ is surjective. In fact, for $f \in f_m = \{f_{m,k}\}_{k=1}^{\infty}$ and for r large enough, there exists $g \in F_{m+r}$ such that

$$\|\phi_{k_m,k_{m+r}}(f) - \pi_{n_{m+r}}^{(k_{m+r})}(g)\| \leq \frac{1}{2^{m+r}}.$$

So

$$\|\phi_{k_m,k_{m+r+s}}(f) - \phi_{k_{m+r},k_{m+r+s}}\big(\pi_{n_{m+r}}^{(k_{m+r})}(g)\big)\| \leq \frac{1}{2^{m+r}}$$

for all s. This implies that

$$\|\phi_{k_m,\infty}(f) - \phi\big(\psi_{m+r,\infty}(g)\big)\| \leq \frac{1}{2^{m+r-2}}.$$

So $\phi_{k_m,\infty}(f)$ must be in $\phi(B)$, the closed range of ϕ. Hence $\phi(B)$ contains the images of $\{f_m\}_{m=1}^{\infty}$ which are dense in A.

We complete the proof by showing ϕ to be injective. Let $b \in B$ be such that $\|b\| = 1$ and $\phi(b) = 0$. We will derive a contradiction. Given $\varepsilon > 0$, there exists some integer m and $a \in A_{n_m}^{(k_m)}$ such that $\|\psi_{m,\infty}(a) - b\| < \varepsilon$. Hence

$$\|\phi\big(\psi_{m,\infty}(a)\big)\| < \varepsilon.$$

Notice that $\|\psi_{m,\infty}(a)\|$ must be no less than $1-\varepsilon$. In fact, for any positive integer r, we have $\|\psi_{m+r,\infty}(\psi_{m,m+r}(a))\| \geq 1-\varepsilon$ and $\|\phi(\psi_{m+r,\infty}(\psi_{m,m+r}(a)))\| < \varepsilon$. Let $a_{m,k} \in a_m$ be such that $\|a_{m,k} - a\| < \varepsilon$. By increasing m we may assume that $a_{m,k}$ is in F_m. Now we have

$$\|\psi_{m+r,\infty}(\psi_{m,m+r}(a_{m,k}))\| \geq 1-\varepsilon \quad \text{and} \quad \|\phi(\psi_{m+r,\infty}(\psi_{m,m+r}(a_{m,k})))\| < 2\varepsilon.$$

So we may assume that a is in F_m.

Consider the diagram we obtained before

$$
\begin{array}{ccccccc}
A_{n_m}^{(k_m)} & \longrightarrow & A_{n_{m+r}}^{(k_{m+r})} & \longrightarrow & \cdots & \longrightarrow & B \\
\downarrow & & \downarrow & & & & \\
A^{(k_m)} & \longrightarrow & A^{(k_{m+r})} & \longrightarrow & \cdots & \longrightarrow & A.
\end{array}
$$

Then

$$\left\|\phi(\psi_{m,\infty}(a)) - \phi_{k_m,\infty}\left(\pi_{n_m}^{(k_m)}(a)\right)\right\| < \sum_{j=m}^{\infty} 2^{-j}$$

or

$$\left\|\phi_{k_m,\infty}\left(\pi_{n_m}^{(k_m)}(a)\right)\right\| < \varepsilon + 2^{-m+2}.$$

Similarly,

$$\left\|\phi_{k_{m+r},\infty}\left(\pi_{n_{m+r}}^{(k_{m+r})}(\psi_{m,m+r}(a))\right)\right\| < \varepsilon + 2^{-m-r+2}$$

for $r = 1, 2, 3, \ldots$.

Denote $\psi_{m,m+r}(a)$ by $\tilde{a}$. Let $\|\tilde{a}\| = \|\tilde{a}(x_0)\|$ for some point x_0 in the spectrum X of a summand of $A_{n_{m+k}}^{(k_{m+k})}$. At this point, $\psi_{m,m+r}$ has the following expression:

$$\psi_{m,m+r}(a)(x_0) \;=\; u \begin{bmatrix} a(t_1) & & & \\ & a(t_2) & & \\ & & \ddots & \\ & & & a(t_\ell) \end{bmatrix} u^*,$$

where $\{t_i\}_{i=1}^{\ell}$ is a group of points in the spectrum of $A_{n_m}^{(k_m)}$. Here $A_{n_{m+r}}^{(k_{m+r})}$ is a finite direct sum of basic building blocks; we write it as a block diagonal form. For another point in X, we have another group of points $\{t_i'\}_{i=1}^{\ell}$. The two groups of points can be paired to within $\frac{1}{2^{m+r}}$ one by one. If r is large enough, we may assume that $\left|\|\tilde{a}(x_0)\| - \|\tilde{a}(x)\|\right| < \varepsilon$ for all x in X. Recall that in 9.3 we showed that at least one point in X has preimage in the spectrum of $A^{(k_{m+r})}$. Hence,

$$\left\|\pi_{n_{m+r}}^{(k_{m+r})}(\tilde{a})\right\| \geq \|\tilde{a}(x_0)\| - \varepsilon.$$

Notice that since $\|\psi_{m+r,\infty}(\tilde{a})\| \geq 1-2\varepsilon$,

$$\left\|\pi_{n_{m+r}}^{(k_{m+r})}(\tilde{a})\right\| \geq 1-3\varepsilon.$$

Since $\|\pi_{n_{m+r}}^{(k_m+r)}(\tilde{a}) - \phi_{k_m,k_{m+r}}(\pi_{n_m}^{(k_m)}(a))\|$ is less than $\frac{1}{2^{m-1}}$, we have that $\|\phi_{k_m,k_{m+r}}(\pi_{n_m}^{(k_m)}(a))\|$ is no less than $1 - 3\varepsilon - \frac{1}{2^{m-1}}$ for all r. Hence $\|\phi_{k_m,\infty}(\pi_{n_m}^{(k_m)})(a)\|$ is no less than $1 - 3\varepsilon - \frac{1}{2^{m-1}}$. Replacing a by $\psi_{m,m+r}(a)$, we have

$$\left\|\phi_{k_{m+r},\infty}\left(\pi_{n_{m+r}}^{(k_{m+r})}(\psi_{m,m+r}(a))\right)\right\| \geq 1 - 3\varepsilon - \frac{1}{2^{m-1}},$$

which says that

$$\|\phi(\psi_{m,\infty}(a))\| \geq 1 - 3\varepsilon - \frac{1}{2^{m-1}}.$$

This is incompatible with the inequality

$$\|\phi(\psi_{m,\infty}(a))\| < \varepsilon$$

obtained before, if ε is small enough and m is large enough.

REFERENCES

1. R. Bhatia, *Perturbation Bounds for Matrix Eigenvalues*, Pitman Research Notes in Mathematics 162, Longman, London, 1987.

2. B. Blackadar, *K-Theory for Operator Algebras*, Mathematical Sciences Research Institute Publications 5, Springer-Verlag, New York, 1986.

3. B. Blackadar, Symmetries of the CAR algebra, *Ann. of Math.* **131** (1990) 589–623.

4. B. Blackadar, O. Bratteli, G.A. Elliott, and A. Kumjian, Reduction of real rank in inductive limits of C*-algebras, *Math. Ann.*, **292**, 111–126 (1992).

5. O. Bratteli, G.A. Elliott, D.E. Evans, and A. Kishimoto, Finite group actions on AF algebras obtained by folding the interval, *K-Theory*, to appear.

6. O. Bratteli, A. Kishimoto, Non-commutative spheres III: Irrational rotations, *Comm. Math. Phys.* **147** (1992) 605-624.

7. L.G. Brown and G.K. Pedersen, C*-algebras of real rank zero, *J. Funct. Anal.* **99** (1991) 131–149.

8. M.-D. Choi and G.A. Elliott, Density of the self-adjoint elements with finite spectrum in an irrational rotation C*-algebra, *Math. Scand.* **67** (1990) 73–86.

9. E. G. Effros, On the structure theory of C*-algebras: some old and new problems, *operator Algebras and Applications* (edited by R. V. Kadison), *Proc. Symp. Pure Math.* **38** (1982), Part 1, pages 19-34.

10. G.A. Elliott, On the classification of inductive limits of sequences of semisimple finite-dimensional algebras, *J. Algebra* **38** (1976) 29–44.

11. G.A. Elliott, On the classification of C*-algebras of real rank zero, *J. Reine Angew. Math.*, to appear.

12. G.A. Elliott, Dimension groups with torsion, *International J. of Math.* **1** (1990) 361–384.

13. G.A. Elliott, A classification of certain simple C*-algebras, *Quantum and Non-commutative Analysis* (editors, H. Araki et al.), Kluwer, to appear.

14. G.A. Elliott, The classification problem for amenable C*-algebras, preprint.

15. G.A. Elliott and D.E. Evans, The structure of the irrational rotation C*-algebra, *Ann. of Math.*, to appear.

16. G. A. Elliott and G. Gong, On inductive limits of matrix algebras over the two-torus, preprint.

17. G. A. Elliott, G. Gong, H. Lin and C. Pasnicu, Abelian C^*-subalgebras of real rank zero and C^*-algebras of inductive limits, preprint.

18. D.E. Evans and A. Kishimoto, Compact group actions on UHF algebras obtained by folding the interval, *J. Funct. Anal.* **98** (1991) 346–360.

19. Halmos, P.R. and H.E. Vaughan, The marriage problem, *American Journal of Mathematics* **72** (1950) 214–215.

20. T. Kato, *Perturbation Theory for Linear Operators*, Springer-Verlag, New York, 1966.

21. A. Kumjian, An involutive automorphism of the Bunce-Deddens algebra, C. R. Math. Rep. Acad. Sci. Canada **10** (1988).

22. T.A. Loring, The noncommutative topology of one-dimensional spaces, *Pacific Journal of Math.* **136** (1989).

23. I. F. Putnam, On the topological stable rank of certain transformation group C^*-algebras, *Ergodic Theory Dynamical Systems* **10** (1990), 197-207.

24. H. Su, K-theoretic classification for certain real rank zero C*-algebras with torsion K_1, preprint.

25. K. Thomsen, Inductive limits of interval algebras: the tracial state space, Amer. J. Math., to appear.

Current Address:

Department of Mathematics
University College of Swansea
Singleton Park
Swansea SA2 8PP
U. K.

Editorial Information

To be published in the *Memoirs*, a paper must be correct, new, nontrivial, and significant. Further, it must be well written and of interest to a substantial number of mathematicians. Piecemeal results, such as an inconclusive step toward an unproved major theorem or a minor variation on a known result, are in general not acceptable for publication. *Transactions* Editors shall solicit and encourage publication of worthy papers. Papers appearing in *Memoirs* are generally longer than those appearing in *Transactions* with which it shares an editorial committee.

As of December 7, 1994, the backlog for this journal was approximately 3 volumes. This estimate is the result of dividing the number of manuscripts for this journal in the Providence office that have not yet gone to the printer on the above date by the average number of monographs per volume over the previous twelve months, reduced by the number of issues published in four months (the time necessary for preparing an issue for the printer). (There are 6 volumes per year, each containing at least 4 numbers.)

A Copyright Transfer Agreement is required before a paper will be published in this journal. By submitting a paper to this journal, authors certify that the manuscript has not been submitted to nor is it under consideration for publication by another journal, conference proceedings, or similar publication.

Information for Authors and Editors

Memoirs are printed by photo-offset from camera copy fully prepared by the author. This means that the finished book will look exactly like the copy submitted.

The paper must contain a *descriptive title* and an *abstract* that summarizes the article in language suitable for workers in the general field (algebra, analysis, etc.). The *descriptive title* should be short, but informative; useless or vague phrases such as "some remarks about" or "concerning" should be avoided. The *abstract* should be at least one complete sentence, and at most 300 words. Included with the footnotes to the paper, there should be the 1991 *Mathematics Subject Classification* representing the primary and secondary subjects of the article. This may be followed by a list of *key words and phrases* describing the subject matter of the article and taken from it. A list of the numbers may be found in the annual index of *Mathematical Reviews*, published with the December issue starting in 1990, as well as from the electronic service e-MATH [**telnet e-MATH.ams.org** (or **telnet 130.44.1.100**). Login and password are **e-math**]. For journal abbreviations used in bibliographies, see the list of serials in the latest *Mathematical Reviews* annual index. When the manuscript is submitted, authors should supply the editor with electronic addresses if available. These will be printed after the postal address at the end of each article.

Electronically prepared manuscripts. The AMS encourages submission of electronically prepared manuscripts in $\mathcal{A}\mathcal{M}\mathcal{S}$-TEX or $\mathcal{A}\mathcal{M}\mathcal{S}$-LATEX because properly prepared electronic manuscripts save the author proofreading time and move more quickly through the production process. To this end, the Society has prepared "preprint" style files, specifically the amsppt style of $\mathcal{A}\mathcal{M}\mathcal{S}$-TEX and the amsart style of $\mathcal{A}\mathcal{M}\mathcal{S}$-LATEX, which will simplify the work of authors and of the production staff. Those authors who make use of these style files from the beginning of the writing process will further reduce their own effort. Electronically submitted manuscripts prepared in plain TeX or LaTeX do not mesh properly with the AMS production systems and cannot, therefore, realize the same kind of expedited processing. Users of plain TeX should have little difficulty learning $\mathcal{A}_{\mathcal{M}}S$-TeX, and LaTeX users will find that $\mathcal{A}_{\mathcal{M}}S$-LaTeX is the same as LaTeX with additional commands to simplify the typesetting of mathematics.

Guidelines for Preparing Electronic Manuscripts provides additional assistance and is available for use with either $\mathcal{A}_{\mathcal{M}}S$-TeX or $\mathcal{A}_{\mathcal{M}}S$-LaTeX. Authors with FTP access may obtain *Guidelines* from the Society's Internet node e-MATH.ams.org (130.44.1.100). For those without FTP access *Guidelines* can be obtained free of charge from the e-mail address guide-elec@ math.ams.org (Internet) or from the Customer Services Department, American Mathematical Society, P.O. Box 6248, Providence, RI 02940-6248. When requesting *Guidelines*, please specify which version you want.

At the time of submission, authors should indicate if the paper has been prepared using $\mathcal{A}_{\mathcal{M}}S$-TeX or $\mathcal{A}_{\mathcal{M}}S$-LaTeX. The *Manual for Authors of Mathematical Papers* should be consulted for symbols and style conventions. The *Manual* may be obtained free of charge from the e-mail address cust-serv@math.ams.org or from the Customer Services Department, American Mathematical Society, P.O. Box 6248, Providence, RI 02940-6248. The Providence office should be supplied with a manuscript that corresponds to the electronic file being submitted.

Electronic manuscripts should be sent to the Providence office immediately after the paper has been accepted for publication. They can be sent via e-mail to pub-submit@math.ams.org (Internet) or on diskettes to the Publications Department, American Mathematical Society, P.O. Box 6248, Providence, RI 02940-6248. When submitting electronic manuscripts please be sure to include a message indicating in which publication the paper has been accepted.

Two copies of the paper should be sent directly to the appropriate Editor and the author should keep one copy. The *Guide for Authors of Memoirs* gives detailed information on preparing papers for *Memoirs* and may be obtained free of charge from the Editorial Department, American Mathematical Society, P.O. Box 6248, Providence, RI 02940-6248. For papers not prepared electronically, model paper may also be obtained free of charge from the Editorial Department.

Any inquiries concerning a paper that has been accepted for publication should be sent directly to the Editorial Department, American Mathematical Society, P.O. Box 6248, Providence, RI 02940-6248.

Recent Titles in This Series

(*Continued from the front of this publication*)

(See the AMS catalog for earlier titles)